LIGHTNING
AND ITS SPECTRUM

LIGHTNING AND ITS SPECTRUM

AN ATLAS OF PHOTOGRAPHS

LEON E. SALANAVE

THE UNIVERSITY OF ARIZONA PRESS   Tucson, Arizona

About the Author . . .

LEON E. SALANAVE applied astronomical spectroscopic
techniques to lightning photography, producing hundreds of
lightning spectra with unprecedented detail, while at the
University of Arizona's Institute of Atmospheric Physics. He
had formerly been at the California Academy of Sciences,
where he was involved in designing and building the Morrison
Planetarium. Later he conducted site tests for the National
Astronomical Observatory project that preceded the found-
ing of Kitt Peak National Observatory. In 1971 he became
executive officer of the Astronomical Society of the Pacific.
A contributor to numerous conferences and a consultant for
astronomical optics at the California Academy of Sciences,
he has taught astronomy at the City College of San Francisco
since 1979.

Frontispiece:
Lightning near Vermillion Cliffs, Arizona

*Photograph by Richard F. McGraw, reproduction
courtesy of the Friends of Photography*

THE UNIVERSITY OF ARIZONA PRESS

This book was set in 12/14 point English Times
on Compugraphic EditWriter 7500.

Library of Congress Cataloging in Publication Data

Salanave, Leon E   1917-
    Lightning and its spectrum.

    Bibliography: p.
    Includes index.
    1. Lightning—Atlases. I. Title.
QC966.S37   551.5'63'0222   80-18882
ISBN O-8165-0374-5

To the fond memory of
Edward T. ("Ted") Pierce
(b. 13 May 1916 — d. 22 February 1978)
for his inspiration in a generation of
research on atmospheric electricity,

and

In grateful recognition of
James ("Jim") Hughes
for his unswerving support of research in the atmospheric
sciences, especially of this work on lightning.

CONTENTS

ILLUSTRATIONS

FOREWORD

Leon Salanave joined the Institute of Atmospheric Physics in late 1960 when I was the director. At the time he didn't know much about terrestrial atmospheric processes, but we at the Institute were more interested in well-trained physicists than we were in weathermen, and he had shown the depth and breadth to which his knowledge, especially in the area of optics, could be applied.

Our faith in serendipity was early and well-rewarded when he proposed that the well known astronomical technique of slitless spectroscopy, and photography with high-acuity lenses of long focal length and large aperture, might be beneficially applied to the subject of lightning. Although there is good reason to believe that lightning may very well strike twice in the same place, certainly it does not do it as consistently and precisely as would be required to get any important number of lightning spectra with the more usual slit spectrograph. However, due to the imaginative Salanave application, the Institute accumulated hundreds of excellent quality spectra that have tremendously illuminated the physics of clouds and lightning in ways which otherwise would have been impossible. And now, some twenty years and thirty scientific papers later, we have *Lightning and Its Spectrum: An Atlas of Photographs,* a collection of the finest in scientific lightning photographs of a quality which the most critical eyes will find acceptable for careful study.

Leon Salanave's ten years of dedicated research in the area he pioneered eventually gave us such milestones as the resolution of fine detail in the infrared, visible, and ultraviolet portions of the spectra — not of just the total flash but of the component strokes; the identification of many excited ionic species, made possible by such detailed spectra; and knowledge of the physical properties of the lightning channel, including temperature, pressure, density, particle distribution, and conductivity. Although his research has already found application in discrimination and detection techniques and the detection and counting of lightning strokes in daylight, surely additional application will be found for the fundamental knowledge on lightning that his program has accumulated.

Early in 1971, Leon Salanave left the Institute of Atmospheric Physics to return to the area of his original professional interests, as executive officer of the Astronomical Society of the Pacific. All of us of the Institute missed his friendly and scholarly companionship and eager willingness to help other colleagues, no matter how hard pressed he was for time for his own work. Some of us seriously doubted that the lightning atlas would ever be completed. However, those individuals of little faith had not read the lesson of persistence which was so well demonstrated in the previous ten years of painstaking research.

We now have the atlas. It is not only a meritorious publication but also a tribute to a fine and imaginative effort. It will be appreciated by its author's colleagues, who will look forward to future contributions.

A. RICHARD KASSANDER, JR.
Vice President for Research
University of Arizona

A WORD FROM THE AUTHOR

My first thoughts on putting together a photographic atlas of lightning and its spectrum came some time in 1969 or 1970, after I had spent a decade with the Institute of Atmospheric Physics at the University of Arizona observing lightning and publishing papers on its optical properties. Privileged as I was to have an extraordinary opportunity to watch the magnificent "flights of thunderbolts" so characteristic of summer storms in the wide open spaces around Tucson, I gradually came to appreciate the fact that many students of the atmospheric sciences in general, and of atmospheric electricity in particular, had never seen the fine detail in lightning — much less had the opportunity of photographing it — as clearly as my colleagues and I in that southern Arizona desert environment. Additionally, and to make it more difficult for such persons to appreciate the details, reproductions of photographs generally to be found in the literature were almost uniformly of mediocre quality, certainly not up to the high standards of modern photolithography and press-work. Typically, much of the fine structure and subtle shading found on original negatives and prints of lightning flashes and their spectra was lost in the halftone prints, but at the same time these invisible details were referred to and discussed in otherwise excellent research papers and review articles.

Over the years, I have received inquiries which suggested that the questioner had never seen photographs or halftone prints of sufficiently high quality to show clearly some particular aspect of lightning's intricate and varied structure. This collection of photographs is intended to remedy that situation. With the passage of time, several well-illustrated books about lightning have been published. Also, the standards of halftone reproduction in the relevant scientific journals have improved. But the *primum mobile* for an atlas of lightning still seems to be in force.

Some of the best photographs in this collection are being published for the first time. Others have appeared elsewhere, usually in connection with articles or reviews cited in the corresponding legends; this was done for the benefit of those readers wishing to pursue details that may not have been clearly reproduced in the original publication. A very few have been adequately reproduced in the cited literature, but they are included in this atlas for the sake of completeness, or because the photograph uniquely shows a particular aspect of the form and structure of a lightning flash.

The legends for the figures are more than labels; they are brief descriptions intended to give sufficient information to make the content of the photograph clear, without going into great detail, and especially avoiding matters which are in the realm of speculation or theory. Thus, as time passes, quantitative data and theoretical interpretations will change, but these photographs and accompanying text can be expected to remain for many years a useful presentation of optical phenomena associated with lightning.

I have compiled a brief glossary for the benefit of readers (especially the browsers!) who are not already familiar with lightning and the terminology associated with its optical study. The bibliography includes a reading list, admittedly very selective, in which I have arranged titles according to various levels — popular, technical, etc. — again for the benefit of those readers not already familiar with relevant concepts and terminology, or for those who want to extend an already working knowledge of basic physics into the special realm of lightning studies.

I am indebted to many persons for providing photographs, information, and advice on the contents of this atlas. The list of photograph credits gives specific acknowledgment of the source for each photograph. However, special mention is appropriate in a few cases.

My thanks go to Prof. Marx Brook at the New Mexico Institute of Mining and Technology, in Socorro, who gave me access to the extensive collection of time-streaked photographs of multistroked flashes and of the continuing current luminosity phenomenon. He loaned me many of his original negatives from which to make reproduction prints and, I must acknowledge, kindly allowed me to keep these irreplaceable data for many months while I was organizing my material.

From the Swiss High Voltage Commission in Zurich, Prof. Karl Berger, Ing. Ernst Vogelsanger, and Mr. Hugo Binz generously provided me with special prints of several of the unique time-streaked photographs of leader and return stroke events, as recorded at the research station on Mount San Salvatore near Lake Lugano, on the Swiss-Italian border. From the beginning, they were enthusiastic about the plans for this atlas and made some of their best material available to me. Also, they were cordial hosts on two occasions (in 1964 and 1967) when I set up my spectrographic camera near Lugano in order to observe lightning flashes over Mount San Salvatore.

Among the most unusual and rare photographs in the atlas are those of triggered lightning (2.17, 2.19, and 3.11), multiple channels (7.1), bead lightning (3.7), and electrical discharges associated with volcanic activity (3.9 and 3.10). Special thanks are due the following persons, respectively, for providing those photographs: Donald Arabian, James Stahmann, Frank Berry and William Regan, Juoko Wesantera, James Hughes, and Sigurgeir Jónasson. Some of the finest prints, especially for Chapter 2, were contributed by George Marcek.

The instrumentation and photographic work on lightning at the Institute of Atmospheric Physics was principally supported by the Office of Naval Research; a grant from the National Science Foundation funded the construction of the lightning observatory. James Hughes, at ONR, always seemed able to find the money to take care of our needs and, together with A. Richard Kassander, then director of the Institute of Atmospheric Physics, saw to it that I never lacked for equipment, materials, personnel, and other resources with which to carry out the original research and subsequent publication of results, including this atlas. I am indebted to Dean John Hensill and professors Clarence Rainwater and Charles Burleson, in the Physical Sciences Division at San Francisco State University, who made working space and some other facilities available during the time I was finishing the collection of material and writing after I left the University of Arizona.

Maury Giles, scientific photographer at the California Academy of Sciences, San Francisco, made several prints that required special care, and George Kew, staff photographer for the Optical Sciences Center at the University of Arizona, was particularly helpful in solving special problems of photo printing for several difficult cases. In addition I would like to thank the staff of the University of Arizona Press for effecting publication, noting in particular my appreciation for the efforts of Marshall Townsend, Elizabeth Shaw, Francis Morgan, Don Pollen, and Kimberley Vivier. Penny Salanave, my wife through the gestation and birth of this work, patiently put up with my labor pains for nearly ten years.

Finally, for all phases of this work — planning, acquiring and writing about the photographs — my special thanks go to Louis Battan and Philip Krider at the Institute of Atmospheric Physics of the University of Arizona, and to Richard Orville in the Department of Atmospheric Science, State University of New York at Albany. It gives me particular satisfaction to recall that Dr. Krider and Dr. Orville first became involved in my ''Lightning Project'' when they were graduate students in Tucson. Dr. Krider critically read and edited all the legends in this atlas. However, any residual errors of commission and omission are my own.

LEON E. SALANAVE

"Nexus"

Painting by Geoffrey Chandler

Lightning and thunder have been regarded with wonder, fear, and superstition since time immemorial, and, to a significant extent, such reactions still persist. Geoffrey Chandler's painting of the awesome spectacle of a dazzling lightning flash coming out of a thunderstorm over the desert is an impressive representation of this display of nature's power. To the wonder of the casual observer must be added the fascination and challenge that thunderstorms hold for the plasma physicist, the atmospheric scientist, and other specialists whose professional interests draw them into the study of lightning and related phenomena. Puzzles abound in even the most recent research on this most astonishing manifestation of atmospheric electricity, but, as explained in A Word From the Author, this atlas focuses on those aspects of the subject which are relatively well known and widely accepted.

The literature on the optical properties of lightning is extensive but not vast (in contrast to its electromagnetic properties, for which the literature is enormous). Therefore, the serious reader can, by going to a few of the basic publications from the 1960s on, be led rather directly to most of the relevant work on the optical plasma physics of the lightning discharge.

The first published speculations that lightning and thunder might be related to the light and sound given off by sparks from the "electrick fire" date from the early 1700s in England. In those days, what later became known as static electricity was usually generated by rubbing glass with silk, or amber with fur. The concept of an "electric fluid" was developed to explain the differing behavior of bodies charged with either a vitreous or a resinous process. The first condition corresponded to what modern experimenters call "positive" polarity, the second to "negative" polarity. This was explained by some in terms of two kinds of electrification; to others, it was a matter of an excess (positive) or a deficiency (negative) of a unique fluid in or around a body that was said to be "charged." Subsequent experiments, culminating nearly 300 years later in the discovery of electrons, proved the "single fluid" idea to be the correct one. By mid-eighteenth century the Leyden Jar — the earliest form of the modern condenser, or capacitor — had made it possible to collect and store electrical charge in order to produce bigger and more powerful sparks.

A German scholar, Johann Winkler, wrote unequivocally in 1746 that lightning and thunder differed from the laboratory spark only in the relative strength of the electrical discharge involved. But he suggested no experiment by which his conclusion might be demonstrated. That essential step in the progress of science was made by a colonial American, Benjamin Franklin. In 1749 Franklin proposed his "Sentry-Box Experiment," in which a person standing on an insulated stool, near a pointed steel rod extending vertically for several yards, might draw sparks from a passing thundercloud. It was not until three years later, in May 1752, that François D'Alibard, who had heard in France of Franklin's proposal, set up the experiment in an open field near Paris. On the tenth of that month, sparks were indeed drawn from a thundercloud overhead, and thus Franklin's theory was confirmed!

Meanwhile, back in Philadelphia, "the obscure American" had hit upon the idea of using a kite to carry a conductor of electricity (a damp string) up into a thundercloud and thus draw a current to earth, the latter to be demonstrated by sparks jumping from the insulated lower end of the string via a dangling metal key, or by charging up a Leyden Jar. By sheer coincidence, Franklin accomplished this demonstration in June of 1752 — a month or less after D'Alibard's success with the sentry-box experiment. Benjamin Franklin was characteristically magnanimous toward the prior success of his French contemporary and, further illustrating his generous nature after his invention of the lightning rod some years later, stated that "as we enjoy great advantages from the inventions of others, we should be glad of an opportunity to serve others by any invention of ours; and this we should do freely and generously.'' In hindsight we can appreciate the great peril Dr. Franklin put himself into while conducting his experiments with lightning, and rejoice that his long and useful life was spared for nearly four decades after his bold demonstration of the electrical nature of lightning. (Less fortunate, however, was a Professor G. W. Richmann of St. Petersburg, Russia, killed in 1753 while carrying out his version of the sentry-box experiment that had been performed without fatality by D'Alibard in France the previous year.)

Franklin was first of all a practical man, satisfied to know the laws of nature and their applications and not concerned about how the thundercloud generates its power. Cloud physics has meanwhile made great advances, but the generation of thunderstorm electricity still presents many unanswered questions, especially those which ask for quantitative results. This is perhaps best illustrated by the following quotation from a review paper by Moore and Vonnegut (1977):

Despite the advances of our knowledge in recent decades, made possible by the modern tools of aircraft, radar and electronics, the thunderstorm remains a major meteorological mystery. As yet, we have no satisfactory understanding of the mechanisms by which these violent convective overturnings of the atmosphere produce rain, hail, electrical effects and tornadoes. The lack of good information on the convective motions within, about, and beneath the storm that is necessary to understand the growth of cloud and precipitation particles and the movement of the charged particles responsible for electrification continues to be a major barrier. Equally important missing data, that are indispensable to an understanding of cloud electrification, are the population of cloud and precipitation particles in the various regions of the cloud and the sign and magnitude of the electric charges that they carry.

The first photographs of lightning were obtained during the 1880s; some showed the ribbon lightning effect produced by a crosswind, and others resolved the individual strokes of a flash when the camera was moved horizontally while the shutter was held open (see Figs. 4.1 and 4.2 for some modern examples of this technique). The first spectra of lightning were photographed in 1894 and 1901, but visual observations — typically made through direct-vision prisms — date as far back as 1868. The first attempts at spectrum analysis for the determination of physical conditions in and around the lightning discharge (ionization levels, generation of ozone, temperature, electron density, etc.) began in 1941. However, significant progress in obtaining quantitative data from lightning spectra did not begin until around 1960. Fast time-resolved photographs of lightning strokes and their spectra date from 1928 and 1965, respectively. Beginning in the late 1960s very fast, sensitive electronic imaging systems capable of good resolution, combined with microprocessor computing systems able to handle fast inputs of vast amounts of data, brought about great advances in recording and analyzing the optical phenomena associated with lightning. These developments are still in progress.

The photographs in this atlas illustrate many of the highlights from seventy-five years of optical studies of lightning.

Thunderheads over Baboquivari as seen from Kitt Peak

Photo by Leon Salanave

CLOUDS

THE ENVIRONMENT OF LIGHTNING

1

Clouds are the environment of lightning. A precursor of an active thundercloud is the *cumulus congestus*. It typically evolves into a *cumulonimbus,* the upper part of which often develops into the shape of an anvil. Precipitation, in the form of rain at the lower levels and of supercooled water droplets and hail at the higher levels, is involved in the mechanism of cloud electrification that eventually builds up the strong electric fields required to produce flashes of lightning. Thunderclouds may form in relatively isolated masses over land that is heated by the sun and overlaid by a deep layer of moist air. This condition is especially true in summertime over mountains that rise from the desert lowlands or plains of Arizona and New Mexico. Such clouds give rise to what are known as local thunderstorms. In contrast to these, some thunderclouds form along an extensive system called a "squall line," over or ahead of a cold front which wedges itself under a mass of warm, moist air. This condition gives rise to a whole string of thunderstorms — often severe storms — which can extend across land and sea for over 100 kilometers. Such storm systems frequently develop in the central part of

Figure 1.1 Cumulus congestus evolving into cumulonimbus

Figure 1.2 Cumulonimbus with well-developed anvil

the United States during spring and summer and are typically accompanied by devastating tornadoes and crushing hail as well as lightning.

Four examples of isolated thunderclouds (Figs. 1.1-1.4). Figure 1.1 shows what is probably a transition between cumulus congestus and cumulonimbus, the former characterized by a well-defined dome, the latter by rain falling out of its base, frequently surmounted by an anvil. Figure 1.2 shows a cumulonimbus cloud with a well-developed anvil; Figure 1.3 may be regarded as

Figure 1.3 Anvil viewed end-on

having an anvil, but viewed end-on. It is a cumulonimbus if rain is falling out of its base.

Some thunderclouds develop such violent updrafts that their domed tops punch through the tropopause and rise one or two thousand meters into the stratosphere, a region of great stability into which clouds normally do not rise unless they have the momentum that can be imparted by strong upward motion in a thundercloud. Figure 1.4 shows a large cloud with a well-developed dome partially veiled by cirrus. If the dome were to penetrate 1500 meters into the stratosphere, its rise through the top of the troposphere would have to be at a rate of about 30 meters per second. *(Figs. 1.1, 1.3, and 1.4 courtesy of George Marcek, Tucson; Fig. 1.2 courtesy of Institute of Atmospheric Physics, University of Arizona, Tucson.)*

Figure 1.4 Cumulonimbus with well-developed dome

Figure 1.5 Cloud above a volcanic eruption

Volcanic clouds (Fig. 1.5). Volcanic eruption clouds can produce extraordinary displays of lightning. Figure 1.5 shows a view of the cloud of volcanic ash and steam that resulted from the eruption of Surtsey, a new island off the southern coast of Iceland. The photograph was taken on 6 December 1963, three weeks after the onset of volcanic activity on the ocean floor at a depth of 130 meters. By that time a small island had formed where before there was only open sea. Some of the eruptions were accompanied by displays of unusual lightning (see Figs. 3.9, 3.10). The violent interaction of hot lava and sea water produces an electrified cloud that carries positive charge, and the resulting discharges of lightning lower positive charge to earth. This contrasts with a typical thundercloud that carries a predominantly negative charge at its base and lowers negative charge to ground. *(Photo by Sigurgeir Jónasson, Vestmannaeyjar, Iceland.)*

Figure 2.1 Typical lightning flash to ground

Typical summer lightning over Tucson, Arizona
Courtesy of George Radda

LIGHTNING FLASHES

TYPICAL FORMS

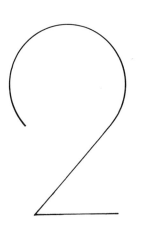

Typical flashes to ground (Figs. 2.1-2.3). Figure 2.1 shows a single flash that was approximately 6 kilometers from the camera. About 3 kilometers of the brilliant path of the discharge channel is visible below the cloud base; a comparable length exists within the cloud, inaccessible to photography. *Branching* is well shown; it is a result of charge being drained from pockets of electrification in the air surrounding the main channel. The great apparent thickness of the luminous path is due to *halation,* a consequence of overexposure on the

Figure 2.2
"Double-grounded" flash

Figure 2.3 Lightning
within cloud and in clear air

original photographic film. The true diameter of the lightning channel is at most a few centimeters.

Figure 2.2 shows a flash making two contacts with the ground, a type of discharge sometimes described as "double grounded." This photograph caught a channel that divides in two at a point about 1500 meters above ground. Actually, as high-speed photography can show in similar cases, the two separate paths to ground develop consecutively rather than branching off simultaneously at the point of division. A faint *stepped leader* (see Fig. 4.6) establishes one path to ground and is followed by a brilliant *return stroke.* Then, in a few milliseconds, a *dart leader* moves smoothly down the path of residual ionization that can exist for as long as 100 milliseconds after the return stroke. It may happen, however, that by the time the dart leader reaches some point on the channel a few hundred meters above the ground, recombinations of ions with electrons will have dropped the conductivity of the air to such a low level that a fresh path will have to be made by a new stepped leader, which takes off from the old channel and makes its separate way to ground. When the resulting flash occurs, it runs up the path established by the new stepped leader to the point where it joins the first channel, and then proceeds along that same path up into the cloud. Thus the picture recorded by a camera has a forked channel, as shown in this illustration. About one quarter of lightning flashes to ground show this effect (see also Fig. 3.5). *(Photos courtesy of Institute of Atmospheric Physics, University of Arizona, Tucson.)*

Figure 2.3 shows lightning within a cloud and in clear air below its base. This kind of display is typical of thunderstorms associated with extensive frontal systems such as occur along the south Atlantic and Gulf coasts of the United States; the photograph was taken from Cocoa Beach on Cape Canaveral, looking toward the

Figure 2.4 Time exposure of desert thunderstorm

ocean. The cloud bases are low — just a few hundred meters above sea level — in sharp contrast to the clouds associated with local thunderstorms in the desert (compare with Fig. 2.4). *(Photo courtesy of George Marcek, Tucson.)*

Time exposures of active thunderstorms (Figs. 2.4-2.5). If the process of charge generation is well developed in an extensive thundercloud, the number of lightning flashes to ground may be as high as three per minute per convective cell. Thus, during a time exposure

Figure 2.5 Time exposure of thunderstorm over mountains

time was about 2 minutes. Note the two fine examples of double-grounded channels on the left (compare with Fig. 2.2).

In the second photograph, Figure 2.5, the camera caught a series of flashes strung out across mountainous terrain near Lake Lugano, Switzerland. Here, the cloud bases are much lower than those over the Arizona desert, and the thunderstorm activity involves a series of convective cells in an extensive weather system. Note the fine example of a double-grounded flash in the center foreground. *Fig. 2.4 courtesy of Institute of Atmospheric Physics, University of Arizona, Tucson; Fig. 2.5 courtesy of Swiss High Voltage Research Committee, Zurich.)*

Lightning striking land and water (Figs. 2.6-2.7). Lightning to ground ordinarily favors high points on the terrain below thunderclouds, such as mountains, hills, and high structures. Otherwise, it randomly strikes to powerlines, houses, large trees, and even open spaces over a large area without prominent relief. Over water, exclusive of projections afforded by ships and smaller craft, there are no obvious preferential points from which upward streamers can rise to make contact with the downward stepping leader. Yet a flash may find a path to the surface of a lake set down among hills and mountains, as Figures 2.6 and 2.7 illustrate.

Figure 2.6 shows Lake Lugano, Switzerland (viewed to the south from the top of Mount San Salvatore), with Mount San Giorgio on the right, both being struck by lightning. Mount San Salvatore is often struck by lightning during intense thunderstorms that form over the mountains south of the Alps, and lightning occasionally strikes the surface of the lake, 640

of only a couple of minutes, about half a dozen flashes may be recorded in a single picture — more if the camera view encompasses more than one active cell.

The first photograph, Figure 2.4, was taken during a violent storm near Tucson, Arizona, and shows a rapid succession of flashes to ground that were generated by a large, active thundercloud. The exposure

Figure 2.6 Lightning striking land and water

meters below the summit of the extremely steep mountain. In Figure 2.7 a double channel is shown just above the surface of the water. It was probably brought about by the junction of two upward streamers that connected with the downward propagating stepped leader. *(Fig. 2.6 courtesy of Hugo Binz, Baden, Switzerland; Fig. 2.7 courtesy of George Marcek, Tucson.)*

Intracloud lightning and air discharges (Figs. 2.8-2.11). Intracloud lightning usually follows a path between the upper, positive and the lower, negative charge centers inside a thundercloud. Thus the luminous channel is typically obscured from view and manifests itself almost entirely in a general illumination of the cloud with an effect sometimes called "sheet lightning."

Figure 2.7 Double
channel striking water

Figure 2.8
Branched air discharge

Obviously, a good clear photograph of this type of lightning would be difficult, if not impossible, to obtain.

Air discharges are lightning flashes that fail to reach the ground. They often follow a heavily branched path in comparatively clear air below the base of a thundercloud and are therefore easily photographed — sometimes with spectacular effects as shown here. Figure 2.8 shows an air discharge over Lake Lugano, Switzerland, with Mount San Salvatore in the foreground. Note the large number of branches, of which one is the brightest and may therefore be considered to

Figure 2.9 Air discharge with fine branching

Figure 2.10 Meandering cloud-to-cloud lightning channel

be the main discharge. In reality, most air discharges are generated by profusely branched stepped leaders, none of which come near enough to the ground to initiate a return stroke. Figure 2.9 shows another air discharge; this one occurred over the desert in southern Arizona.

Compare this with Figure 2.8 and note that here the main discharge channel ends in more fine branching than in the former case. In other respects, the two discharges are very similar. Figure 2.10 shows what appears to be a much less complicated air discharge. It

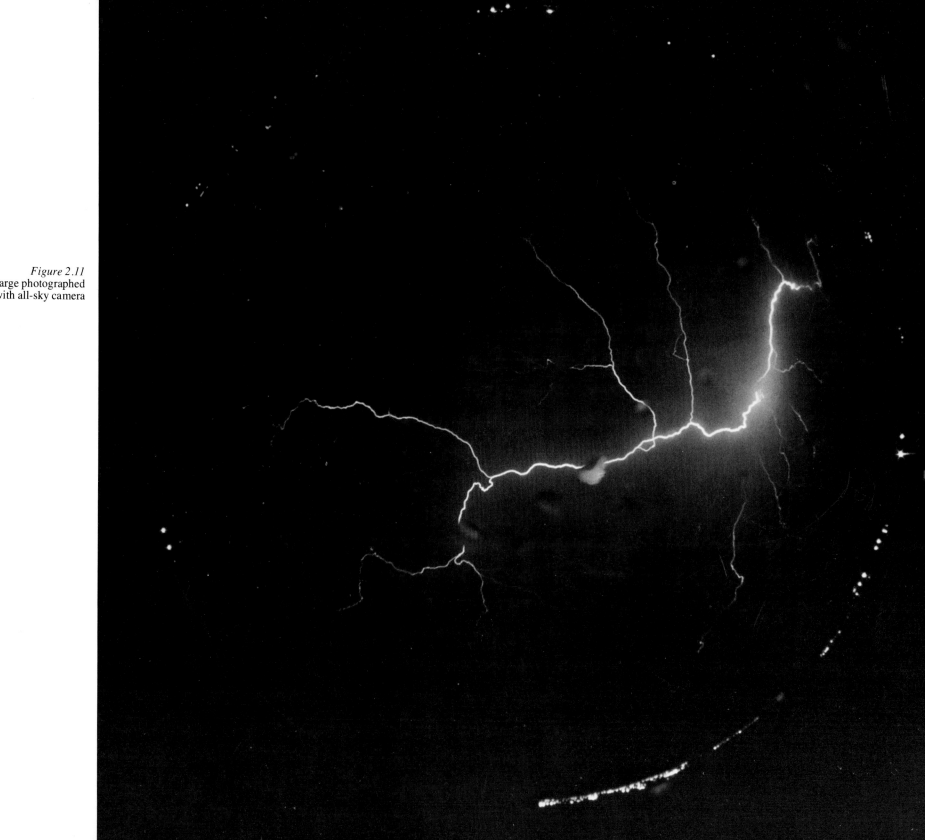

Figure 2.11
Air discharge photographed
with all-sky camera

may be a branch off of a larger flash that happened to be out of the camera field. The tortuous, meandering characteristic of the lightning channel is well illustrated.

Figure 2.11 is a remarkable photograph of an air discharge taken with an all-sky camera — a 220 degree field with the axis of the camera pointed vertically. This reproduction was made from a 35-mm film transparency on Kodachrome II. The lens, a Nikkor "fish-eye" with a focal length of 6 millimeters, was set at aperture stop f/8; note the excellent definition of city lights on the horizon around the observation site, which was in the desert near Tucson, Arizona. The length of the main channel is estimated to be about 16 kilometers, at an average height of 1800 meters above ground — about 300 meters below the prevailing cloud base. This photograph, and information obtained from it, has been described in detail by E. P. Krider (1974). *(Fig. 2.8 courtesy of Swiss High Voltage Research Committee, Zurich; Figs. 2.9-2.11 courtesy of Institute of Atmospheric Physics, University of Arizona, Tucson.)*

Combination of a ground flash and an air discharge (Fig. 2.12). This unusual photograph shows a lightning discharge with two principal branches; one

Figure 2.12
Combination of ground
flash and air discharge

Figure 2.13 Multiple flashes from one thundercloud

(24)

reached the ground, and the other terminated in the air. The main branch point is several hundred meters below the cloud base; note the many lesser branches, particularly along the channel of the air discharge. The sequence in which the two branches developed cannot be determined without a photograph taken with a high-speed moving-film camera. It is unlikely they occurred at the same instant. However, judging from time-resolving photographic observations made of similar events, it is probable the ground flash preceded the air discharge by at least a few milliseconds or as much as a tenth of a second.

Figure 2.12 shows a smearing or broadening of the lightning channel. That this was not due to some vibration of the camera during the exposure is shown by the sharp images of lights in the lower foreground. The effect is called ''ribbon lightning'' and is due to a slight lateral displacement of the discharge channel by a crosswind of sufficient speed to shift the images of the several consecutive strokes within a flash by an appreciable distance on the film (see Figs. 3.1, 3.2, and 3.4).

Multiple flashes from one thundercloud (Fig. 2.13) This photograph was taken with the same camera and from the same site as Figure 2.12 (notice the pattern of lights in the center foreground). It vividly illustrates how several flashes of lightning may emerge in quick succession from a relatively small volume, or cell, of electrification inside a thundercloud. A series of such cells can produce impressive displays of lightning like those captured in Figures 2.4 and 2.5.

Both Figures 2.12 and 2.13 were photographed in summer 1956 at Socorro, New Mexico, with an aerial camera adapted for lightning photography. The lens had a focal length of 155 mm; the camera field was 72 degrees. Under a cloud base approximately 3000 meters above ground, these flashes were about 10 kilometers away. The length of the air discharge in Figure 2.12 was estimated to be at least 10 kilometers, longer if the effect of foreshortening were taken into account. *(Photos courtesy of New Mexico Institute of Mining and Technology, Socorro.)*

Lightning to towers and other elevated structures (Figs. 2.14-2.18). Figure 2.14 shows a bright flash to a metal tower near the summit of Mount San Salvatore,

Figure 2.14 Lightning flash to tower

Figure 2.15
Upward flash from tower

above Lake Lugano in Switzerland. The strike did not hit the tip of the steel needle, which is essentially a lightning rod, but made contact with the framework of the tower several meters below the tip of the needle. However, a faint streamer originated at the tip; it is visible coming out in a vertical direction, just above the bright channel. As is usual for a downward flash, there are numerous branches. One, the brightest, disappeared behind the brow of the hill and terminated in the air. That this is the case was shown by a moving-film exposure taken simultaneously with this photograph. This film, discussed in some detail by Berger (1967: pp. 505, 508), showed that the flash was composed of three component strokes. The bright branch appeared only with the first stroke and therefore followed a stepped leader that failed to reach the ground, not developing a return stroke. This typical sequence of events is illustrated in Figures 4.1 and 4.8. Less commonly, a flash may touch ground at two points, but in that case it is a combination of two consecutive strokes, in which the second stroke departs from the channel established by the first at some point below the cloud base, as illustrated in Figure 2.2.

The sheath of light, or halo, along the lightning channel can probably be explained by an effect of halation in the photographic film, diffusion of light in rain or mist between the channel and the camera, or both. Some speculations have been advanced to support the idea that a discharge corona can sometimes produce an extended glow around the lightning channel, but this idea is not generally accepted. In any case, it would be difficult to rule out from this photograph alone the more obvious explanation of rain and mist around the lightning (see Fig. 7.1 for a similar halo effect, which is considered in still more detail in the discussion of Figs. 7.4-7.6). *(Photo courtesy of Swiss High Voltage Research Committee, Zurich.)*

High structures made of steel, well grounded, and topped with a metallic rod frequently initiate an upward discharge that branches toward the base of the thundercloud, as shown in Figure 2.15. Each branch follows the path established by a stepped leader, just as in the case of downward stepping leaders from a cloud toward the ground (see Figs. 4.6 and 4.13). In some circumstances, one of the upward branches will provide a path for subsequent combinations of a dart leader and return stroke, which are more usually associated with cloud-to-ground flashes. This photograph, together with Figure 2.16, is one of the most spectacular of many that have been obtained at or near the two towers on Mount San Salvatore, Switzerland.

Earlier studies of lightning with moving-film photography and electrical measurements were made by observing flashes to the Empire State Building in New York City. Beginning in 1936, K. B. McEachron and his associates at the General Electric Company discovered some of the characteristics that distinguish between downward flashes from clouds and upward flashes from high structures (see McEachron, 1941; Hagenguth and Anderson, 1952). This type of investigation was greatly enlarged upon and advanced by K. Berger and his associates in the Swiss High Voltage Research Committee. Berger (1967) gives many detailed references to these and related observations. *(Photo courtesy of R. E. Orville, State University of New York at Albany.)*

Loop-the-loop lightning! The unusual flash in Figure 2.16 was initiated by an upward propagating leader from a tower with negative polarity. It looped around to within 600 meters of the tower and then continued along a nearly horizontal path for approximately 2 kilometers before moving out of the camera's field of view. This photograph was taken with a 50-mm lens on 35-mm film, from a distance of 6.5 kilometers, so the trajectory in a horizontal plane with its origin at the tower could be approximately plotted (see Orville and Berger, 1973). Simultaneous electrical measurements at the research station atop Mount San Salvatore showed that current flowed for at least 100 milliseconds but did not exceed 1600 amperes at any time. The positive charge transferred to ground (electrons moving upward effectively transfer positive charge downward) was estimated to be between 30 and 40 coulombs.

This flash could have been triggered by a high ambient electrostatic field at the tower, or by a high transient field induced by a heavy discharge — possibly involving a peak current of many tens of thousands of amperes — at some distance from the tower. However, the set of observations made at the time of this remarkable event cannot resolve the question. The for-

Figure 2.16 Looped upward flash from tower

mation of low stratus clouds around the mountain may explain the unusual horizontal direction of this flash, which is not representative of lightning initiated by upward propagating leaders as is, for example, the upward branching flash pictured in Fig. 2.15. *(Photo courtesy of Richard E. Orville, State University of New York at Albany.)*

Figure 2.17 shows an example of lightning hazard to launching spacecraft. On 14 November 1969, shortly after 11:22 A.M. and 36 seconds after liftoff, the Apollo 12 spacecraft and its Saturn V rocket were struck by lightning. The vehicle was nearly 2000 meters above ground when observers saw a brilliant flash between the rocket and its launching pad. Figure 2.17 is from a photograph of this awesome event. It is an enlargement made from one frame of a 16-mm motion picture film that was started at liftoff. The camera was 400 meters from the launching tower, which is seen silhouetted against clouds illuminated by lightning only 30 meters away. This stroke occurred 60 milliseconds after another, more distant one that was off to the right approximately 500 meters and not recorded by the camera; both strokes lasted about 50 milliseconds.

The crew inside the command module reported numerous electrical disturbances in their equipment, but no irreparable damage to vital systems or components occurred. Astronauts Charles Conrad, Alan Bean, and Richard Gordon were able to restore all circuits to operation and to reset the vital guidance system to its proper alignment. The second manned landing on the moon was distinguished by pinpoint accuracy, but the lightning event, which could have aborted the mission due to serious systems failure or even greater tragedy,

Figure 2.17 Lightning striking Apollo 12

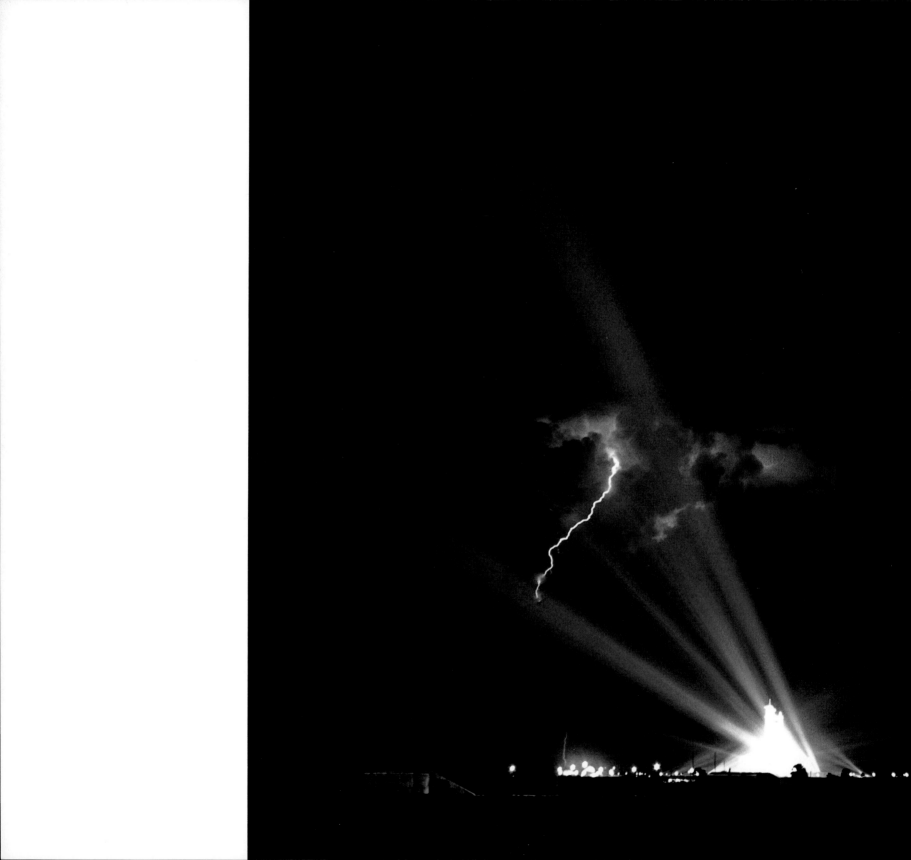

alerted scientists and engineers to the dangers of triggered lightning in the environment of large airframes and in the exhausts from rockets and jet engines.

The lightning to Apollo 12 was presumably triggered by the injection of the metallic rocket and its partially ionized exhaust plume into an electrified environment that would not have produced lightning in and of itself. For 6 hours prior to launch time, and for a similar period afterward, no other lightning strokes were observed or recorded, and no thunder was heard in the vicinity of the launch site. Apollo 12 was launched during a time when a cold front was passing over Cape Canaveral. The flash to Apollo 12 probably followed a leader initiated by the rocket and in that respect was somewhat similar to a flash to a high tower (see Fig. 2.15), or even more closely resembled the flash to an eruptive water plume (see Fig. 3.6).

Figure 2.18*a* is a nighttime scene at Cape Canaveral, photographed on an occasion when a thunderstorm was in the vicinity of the spacecraft launching area. Floodlights illuminate Apollo 17, its Saturn V rocket, and connected umbilical tower; lightning plays in the background. Figure 2.18*b*: A closeup of Apollo 15 atop its launching rocket, attached to the tower by an umbilical cord, with a lightning flash in the distance. Clearly, lightning presents very real hazards to the tall, elaborately instrumented launch towers and the rockets connected to them. Monitoring the level of atmospheric electricity and providing lightning protection around these sites have become matters of high priority at the Kennedy Space Center. (*Figs. 2.17, 2.18b courtesy of National Aeronautics and Space Administration; Fig. 2.18a by George Marcek, courtesy of University of Arizona, Tucson.*)

Figure 2.18a (at left) Lightning near Apollo 17

Figure 2.18b (below) Lightning near Apollo 15

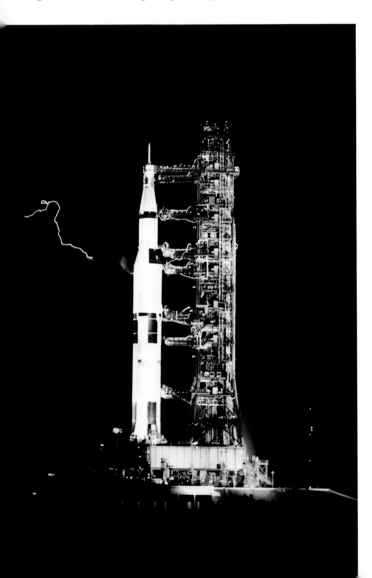

Figure 2.19 Lightning flash intentionally triggered

Lightning flash intentionally triggered by a rocket carrying a thin wire aloft (Fig. 2.19). This photograph was taken aboard the research vessel *Thunderbolt,* cruising under thunderstorms off the west coast of Florida in August 1966. When electric field meters aboard ship indicated the onset of field strengths high enough for the occurrence of lightning, a small rocket which trailed 100 meters of 0.2-mm diameter steel wire was fired into the cloud and photographed at a distance of 10 meters. The spiral contour of the wire is shown in profile at the right edge of the flash. Gradually decreasing ionization from the vaporized wire is followed by a succession of 10 fainter strokes as the channel drifts to the left, carried along by a crosswind at approximately 6 meters per second. The vertical section of the broadened channel in the photograph is nearly 6 meters long, and the entire section at the top of the frame is about 1 meter across.

Out of 23 firings, 17 flashes were brought down to a grounded terminal and passed through a shunt for measuring current and charge transfer. The flash pictured here peaked at 40 kiloamperes and transferred approximately 1 coulomb of negative charge. Insofar as a discharge was triggered in a situation where lightning might not have otherwise occurred, this event was similar to those shown in Figures 2.17 and 3.6-3.9.

The camera was a Polaroid, loaded with Type 55 film. Its 125-mm lens was stopped down to f/16 and capped with a 10 percent transmission filter to prevent overexposure. The shutter was opened 10 seconds *after* the rocket was fired. Techniques used aboard the *Thunderbolt,* and some of the results obtained, have been described by Newman (1969) and Newman et al. (1967). *(Photo courtesy of Lightning and Transients Research Institute, Miami.)*

Lightning photography with a small camera (Figs. 2.20-2.22). Amateur photographers can obtain interesting and satisfying pictures of lightning, using normal lenses and cameras that carry 35-mm or 2¼ ″ x 2¼ ″ (#120) roll film. Fine-grain, medium-speed, black-and-white film can give sharp negatives which will stand enlargement up to ten times and still reproduce the intricate detail of lightning's structure. Shooting in color is nice but gets expensive and is really not necessary in order to do the subject justice. Some photographers even prefer black and white for its greater latitude of exposure and its versatility in developing and printing.

All one need do is mount the camera on a tripod and point it toward an active nighttime thunderstorm. Set the lens stop at F/8 or F/11, with the exposure control on ''B'' or ''T'' for open shutter operation. After at least one good, clear flash has struck in the camera's field of view, open the shutter and wait; another flash will probably appear in that same general direction within a minute or two. If that happens, close the shutter and advance the film. On the other hand, if the clouds are brightly illuminated several times by flashes outside the field of view, advance the film and start another exposure. Otherwise, the film may be badly preexposed before it can record a flash. Keep up this procedure, turning the camera from time to time and pointing it in the direction of ''the action.''

The three photographs shown here were taken in Tucson, Arizona. They are good examples of the excellent results that can be obtained when care is taken in all steps of the process, from selecting the film to making the enlargement. Technical data are as follows. Figure 2.20: Pentax with Asahi Takumar F/2, 55-mm lens, Ilford Pan F film, 5x enlargement; Figure 2.21:

Figure 2.20 Lightning photographed with Pentax camera

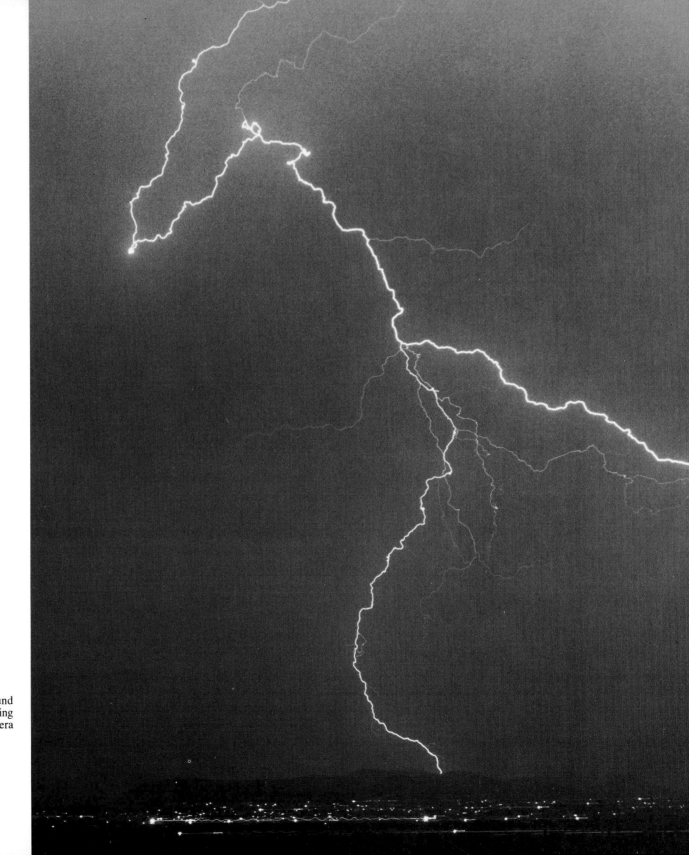

Figure 2.21 Ground
flash with extensive branching
photographed with Rolleicord camera

Figure 2.22
Air discharge photographed
with Rolleicord camera

Rolleicord with Zeiss Planar F/3.5, 75-mm lens, Kodak Plus X film, 4x enlargement; Figure 2.22: Rolleicord with Schneider Xenar F/3.5, 75-mm lens, Kodak Panatomic X film, 7x enlargement. Agfa F and Adox KB-14 films also give good results. A soft working developer, Kodak D-23 or a similar metol formula, gives best results with all the types of film indicated. *(Photos courtesy of George Marcek, Tucson.)*

(35)

Lightning over Surtsey's eruption in December, 1963

Photograph by Sigurgeir Jónasson

Figure 3.1 Ribbon lightning

LIGHTNING FLASHES

UNUSUAL EVENTS

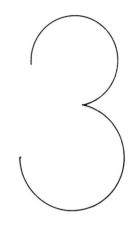

Ribbon lightning (Figs. 3.1-3.4). This picturesque term, popularly applied when the lightning channel to ground appears to be a luminous band rather than a narrow line, suggests that the image of the discharge path has somehow been broadened. The effect arises from two circumstances that must occur together: image or object motion, and an extended duration of the flash. The first can be due to a rapid horizontal eye movement, camera vibration, or transport of the lightning channel by a crosswind. The second circumstance arises when the lightning flash consists of several discrete strokes — possibly with some continuing current luminosity between them — all spanning a time interval of a few hundredths, or even tenths, of a second. Figures 4.1-4.5 show some clearly resolved photographs of multi-stroked flashes. With reduced resolution or smearing together of the individual strokes, these too would appear as "ribbon lightning."

The first photograph of ribbon lightning was probably one taken by H. Kayser, a German physicist, in

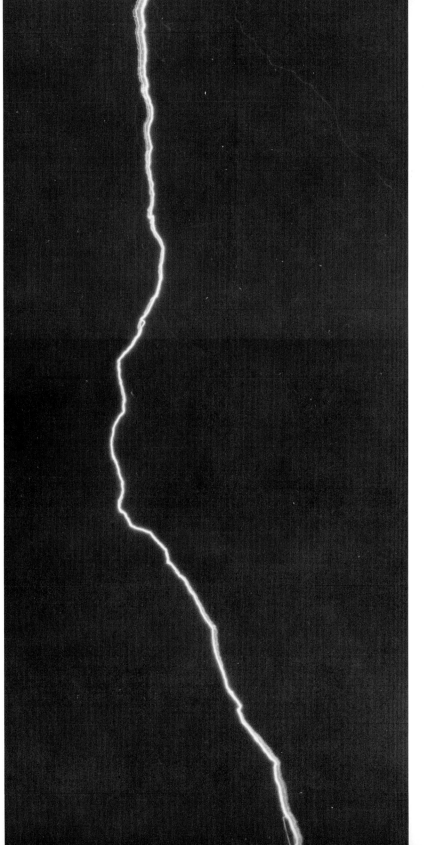

Figure 3.2 Ribbon
lightning showing effect
of reversed wind direction
across channel

1884. He correctly attributed his result to the effect of a
crosswind and inferred that a lightning flash may consist of several distinct discharges occuring over an interval of several tenths of a second or longer.

Three examples of ribbon lightning are shown.
Figure 3.1 was taken with a tripod-mounted camera
whose lens had a focal length of 75 millimeters.
Silhouettes of tree branches in the foreground appear to
be sharp; this tends to rule out camera motion and favor
the effect of a crosswind. However, the possibility of
camera motion cannot be entirely eliminated in this
case. The requirement that the flash consist of several
strokes was clearly satisfied, because at least three
images of the channel can be seen in the photograph.

Figure 3.2 was taken with a massive, sturdily
mounted aerial camera whose telephoto lens had a focal
length of 122 centimeters. In this case it is practically
assured that the camera did not vibrate, and therefore
the ribbon lightning effect was due to a crosswind shift-
ing the images of the strokes during the exposure.
Notice, furthermore, that the direction of the shift — as
shown by the sequence of two faint strokes relative to
the single bright one — is reversed at the top and bottom
of the picture frame, with no perceptible broadening
near the center. The distance of the flash was estimated
to be 4 kilometers; 10 millimeters in the reproduction is
approximately 8 meters at the flash. The vertical section
of the lightning channel shown is approximately 200
meters long and centered about 250 meters above the
ground.

To obtain a rough estimate of the difference in
wind speed required to produce the observed effect, one
can assume a quite reasonable value of 0.15 second for
the total duration of the flash. Then it follows from
measuring the width of the ribbon at the top and bottom

of the frame that the components of wind speed across the line of sight were moving on the order of 30 meters per second relative to each other, or about 15 meters per second relative to the ground at the levels observed. While this represents a strong wind, such gusts sometimes occur near the ground and below the center of an active thundercloud. The most interesting point here is that the gusts were in opposite directions. *(Fig. 3.1 courtesy of George Marcek, Tucson; Fig. 3.2 courtesy of Institute of Atmospheric Physics, University of Arizona, Tucson.)*

Figure 3.3 provides a vivid demonstration that winds just above the ground, and in layers separated by only a few tens of meters, can be blowing in opposite directions. The tower is 135 meters high and, at the time the photograph was taken, smoke plumes were issuing from the 15-, 46-, and 107-meter levels. The trajectories of the plumes were not necessarily perpendicular to the line of sight, but the lowest one had a transverse component opposite in direction from those of the upper two plumes. *(Photo courtesy of Brookhaven National Laboratory, Long Island.)*

Figure 3.3 Smoke plumes aloft showing crosswinds with directions reversed

Figure 3.4, a fine example of ribbon lightning, was photographed in 1959 in the vicinity of Oswestry, near the Welsh-English border. The flash hit a chimney approximately 450 meters from the camera while the shutter was open for only a few seconds. The lens was a Schneider F/6.8 "Angulon," stopped down to f/16 and used with 10 x 12.5-cm sheet film. The photographer's notes on this picture do not give the focal length, but it was probably 90 mm, corresponding to one of three models for this type of lens. The ribbon effect is very clearly shown, broadening toward the top. The original photograph shows seven discrete strokes, with the brightest one on the right. The smoke plume from the chimney drifts to the left, indicating that a component of the wind direction is from the right. The brightest stroke occurred first and was followed by a series of fainter ones, as is usually the case in multistroked flashes (see Fig. 4.1).

The chimney is 26 meters high, but about half of it is hidden behind a building in the foreground. The flue has an outside diameter of 2.7 meters, and evidently the width of the lightning ribbon is greater than that. Measurements made at 65 evenly spaced points on the original photographic print, beginning just above the first sharp bend (60 meters above ground) and ending right below the sharp bend at the top (250 meters above ground), gave apparent horizontal displacements ranging from 3 meters at the bottom to 5.5 meters at the top. It is important to note, however, that these are only apparent shifts, that is, they are the actual displacements viewed as projected onto the plane of the photograph. There is no data on the direction of the wind. It is unlikely that the wind was blowing across the line of sight; its direction as well as its speed may have varied with height. On the assumption that the seven strokes

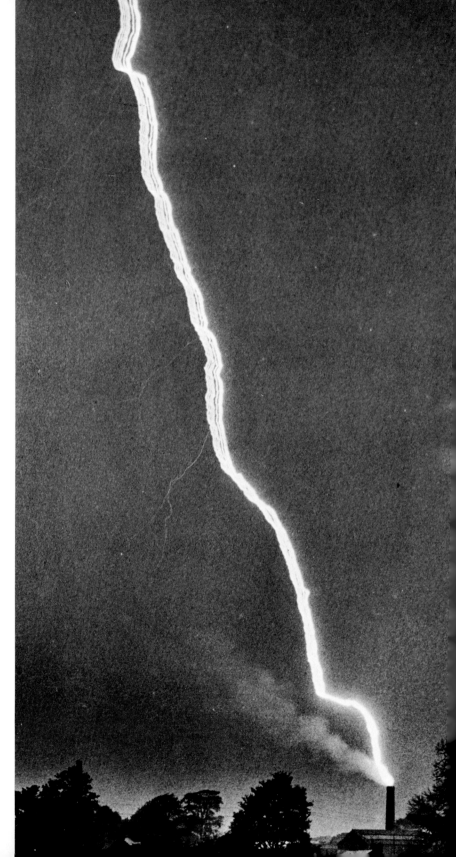

Figure 3.4 Ribbon lightning striking chimney

occurred in an interval of 350 milliseconds, it follows that wind speeds ranged from 9 to 16 meters per second, from the lower to the upper level represented. These numbers are in good agreement with the estimates obtained with Figure 3.2, especially since estimates of the duration of a multistroked flash may be off by a factor of two. True wind speed will be greater than these estimates, on the average, because of the projection factor mentioned above. This photograph has been analyzed in detail by Orville (1977*a*) to obtain a profile of the wind between the 60-meter and 250-meter levels at the time of the lightning flash. *(Photo by T. L. Morgan, reproduction print courtesy of R. H. Golde.)*

Forked lightning flash, with two strokes separated by a crosswind and taking different paths to ground (Fig. 3.5). Figures 2.2, 2.4, and 2.5 all show one or more flashes with apparent double grounds; the sequence of events leading to this characteristic is explained in the caption for Figure 2.2. As far as those photographs show, the lightning channels are all very sharply defined without any broadening, or ribbon, effect. In Figure 3.5, however, the flash has been clearly resolved into two strokes closely paralleling each other in the upper two-thirds of the frame but widely divergent below. (The same massive camera was used here as for Fig. 3.2, so camera motion is practically ruled out as the cause of image separation.)

The distance to the flash was estimated to be 8 kilometers; the scale in this figure (enlarged four times from the original negative) is 10 millimeters = 15 meters. Assuming the interval between strokes is 100 milliseconds, or 0.1 second, measurements of the *horizontal* separation of the channels at several positions give an average displacement of 1.0 millimeter, which translates

Figure 3.5 Forked flash with strokes separated by crosswind

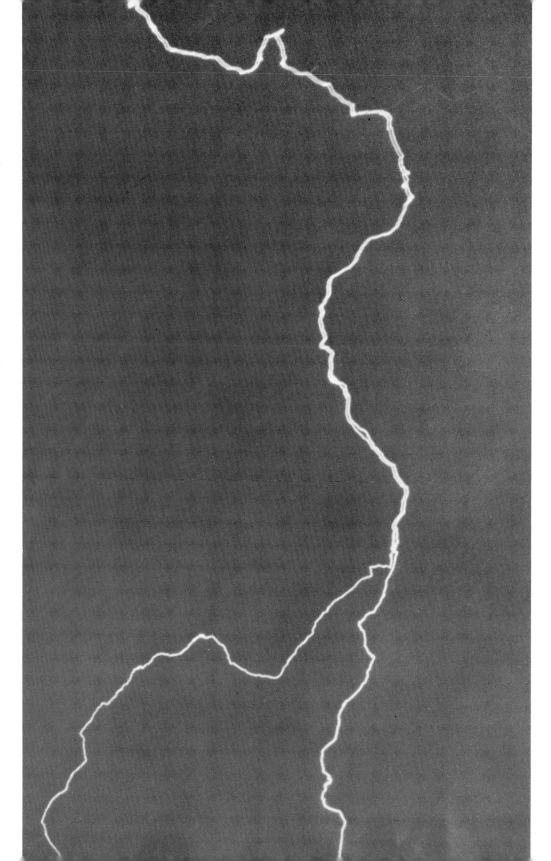

Figure 3.6 Lightning flash triggered by plume from underwater explosion

to 15 meters per second for the speed of the crosswind. This is rather typical of winds near the ground in the vicinity of an active thunderstorm. *(Photo courtesy of Institute of Atmospheric Physics, University of Arizona, Tucson.)*

Bead lightning (Figs. 3.6, 3.7). Bead lightning is a comparatively rare phenomenon, but — unlike ball lightning — there exists photographic evidence that can be accepted in support of numerous reports based on visual observations, however much may be lacking in the way of a satisfactory physical explanation. Hence bead lightning is represented in this atlas while ball lightning is not. In any case, the reader interested in the latter phenomenon should consult Singer (1971).

Figure 3.6 is an enlargement made from one frame of a 35-mm motion picture taken from a distance of 320 meters during the underwater test of a conventional depth charge in Chesapeake Bay on 14 June 1957. A lightning flash composed of four distinct strokes struck the top of the plume between 1.62 and 2.43 seconds after detonation of the depth charge; this is the third stroke. The height of the plume is about 90 meters in this frame, which was exposed 2.39 seconds after detonation.

At the time of detonation, the sky was partially overcast, and a thunderstorm was approaching the test site but still at a distance of 3 or 4 kilometers. It was never the intention of the test crew to detonate the charge in the vicinity of a thunderstorm — quite the contrary! But some difficulty with associated electronic equipment delayed the test for over an hour and, by that time, a thunderstorm had begun to build up near the site. The water plume was inadvertently and abruptly injected into an atmosphere with a large electric field;

Figure 3.7a Bead lightning 15 milliseconds after third stroke to water plume

Figure 3.7b Bead lightning 45 milliseconds after third stroke to water plume

Figure 3.7c Bead lightning 30 milliseconds after fourth stroke to water plume

this triggered the lightning. The same phenomenon sometimes occurs when rockets are fired into, or in the vicinity of, thunderclouds. The most notable such occasion was the one associated with the launching of Apollo 12 (see Fig. 2.17).

Figure 3.7 (*a, b*) shows two frames from a 16-mm motion picture film exposed about 15 and 45 milliseconds after the third stroke to the water plume, respectively. The fading discharge channel has apparently broken up into discrete luminous segments, or beads. Actually these are rather large, isolated sections of the channel; measurements made on copies of the original negative suggest that many of the segments are 2 or 3 *meters* long, as viewed in projection onto a plane perpendicular to the line of sight. However, there seem to be smaller luminous sections that might more aptly be described as beads. Finally, Figure 3.7c is reproduced for a comparison of the beaded structure of the channel at a later time. It shows the fourth and final stroke about 30 milliseconds after its initiation. The plume was nearly 100 meters high at this instant and rising at a rate of approximately 30 meters per second. *(Photos courtesy of U.S. Naval Ordnance Laboratory, Silver Spring, Maryland.)*

Short discussions of observations and theories about bead lightning, together with references to other work on the subject, are given by Uman (1969*a* Appendix B; 1969*b*).

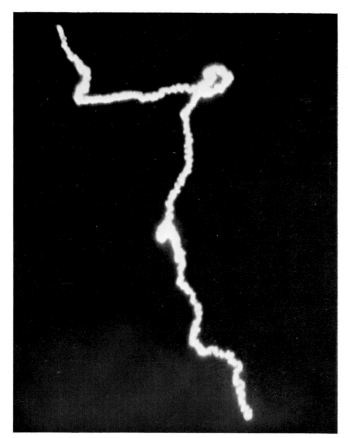

Figure 3.8a Third frame from
film of bead lightning to tower

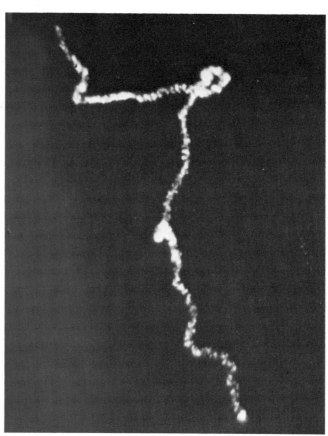

Figure 3.8b Fourth frame from
film of bead lightning to tower

Figure 3.8c Fifth frame from
film of bead lightning to tower

Intricate structure of a lightning channel; three consecutive frames from a high-speed camera apparently showing rapid decay into bright segments (Fig. 3.8). The photographs are from a sequence taken with a Beckman-Whitley "Dynafax" running at 13,000 frames per second. Its 100-mm telephoto lens was focused on, and 8 meters above, the tip of a tower atop Mount Bigelow near Tucson, Arizona. The incandescent tip of the lightning rod is imaged as a bright spot at the bottom of each frame; the rod was 112 meters from the camera.

The photographs are the third, fourth, and fifth frames, respectively, in a series of eight that were recorded on the film. (The first two were grossly overexposed and those following the fifth one were too underexposed for reproduction here.) Note that the interval between frames is not quite 80 *micro*seconds, with an exposure of 2 microseconds per frame. Compare this with the much longer interval of 30 *milli*seconds between exposures shown in Figure 3.7. There is a superficial resemblance in morphology, but the scales — both in time and in distance — are very different. In the photographs of the stroke to the tower, the tortuous section of channel projected onto a plane perpendicular to the line of sight is about 12 meters long. Irregular, luminous features separated by only a few centimeters in projection are resolved. But in the photographs of the fading stroke above the water plume, the apparent length of the channel is nearly 35 meters, and the smallest resolved features are about 75 centimeters long. In either case, however, lengths of the order of tens of centimeters are to be compared with the generally accepted estimate of 1 or 2 centimeters for the diameter of the lightning channel.

Then the question arises as to what should be considered a bead. The configuration is sometimes picturesquely described as a "string of sausages" — and very thin ones at that! The appearance of alternate bright and dark sections along a fading lightning discharge may be due to intrinsic variations in luminosity along its length, or it may result simply from the geometry of the tortuous, meandering channel — or a combination of both. Where a section lies near the line of sight, one looks through a longer column of incandescence than elsewhere. Because the heated gas is optically thin — that is, there is almost no self-absorption of emitted radiation — the greater total path from which light is received will result in an enhanced brightness (a "bead") compared to those sections that are viewed more nearly broadside. As the channel fades, the latter sections will disappear while the former persist.

Explanations calling for intrinsic variations in apparent brightness along the channel, involving either changes in size or in radiance and its duration, generally invoke very special phenomena to modulate these characteristics along the length of the channel in order to produce the observed effect. *Magnetic pinch,* an effect that can be demonstrated in a laboratory and explained by plasma physicists, is one such, but its application to lightning has not been demonstrated satisfactorily. Perhaps as well as any photographs can show, these that record rapid changes in the appearance of a lightning channel hint at what complexities are involved — many of them unexplained. Figures 5.6-5.8 also show detail of this kind, and their respective legends discuss other results obtained. See also Evans and Walker (1963). *(Photos courtesy of Institute of Atmospheric Physics, University of Arizona, Tucson.)*

Unusual lightning from a volcanic cloud (Figs. 3.9, 3.10). During November 1963 a series of eruptions on the ocean floor off the south coast of Iceland formed a new island, since called Surtsey. From its beginning, the process of island-building was accompanied by violent outpourings of gas, steam, and ash. During the fourteenth day of its eruption Surtsey was first reported to be the scene of spectacular lightning discharges, and a few days later an especially vivid display occurred when a thunderstorm apparently gathered over the island.

These two photographs were taken on the evening of December 1, between $19^h 27^m$ and $19^h 48^m$ local time. Figure 3.9 is a 5-minute exposure showing the peculiar form of air discharge that was characteristic of the lightning activity over the island. Lightning over Vesuvius during an eruption in 1944 showed a very similar form. Figure 3.10 is a 10-minute exposure that began only 6 minutes after the preceding exposure ended. Note the heavy discharge to ground which resembles, at least in form, the usual flash to ground observed in ordinary thunderstorms. Both of these photographs were taken when a full moon was above the horizon and to the left of the cloud; this accounts for the general, diffuse illumination on that side of the picture. The camera was located on the Icelandic coast, at Vestmannaeyjar, and pointed toward the eruption at

Figure 3.9 Air discharge from volcanic cloud

a distance of 23 kilometers in a southwesterly direction. *(Photos by Sigurgeir Jónasson, Vestmannaeyjar, Iceland.)*

When clouds of water vapor and finely pulverized rock billow upward they become electrified, usually with a strong positive charge. This is the opposite of the situation existing in a typical thundercloud, which has a predominantly negative charge near the ground. The charging mechanism obviously involves the interaction of particles of lava (tephra) and water (steam), but the details are not clear. It is tempting to suppose that the violent action of molten lava in contact with sea water is a necessary condition for this type of cloud electrification, but such an interface is not the only situation leading to this kind of electrified cloud. Several volcanoes not in contact with the sea have produced displays of lightning: Etna, Vesuvius, Paracutin, and Fernandina, for example. Electrical field measurements made during times of lightning activity over the islands of Surtsey and Heimaey showed that the discharges lowered positive charge to ground. Lightning events associated with eruptions of Surtsey (1963-64) and neighboring Heimaey (1973) have been described and analyzed by Anderson et al. (1965) and Brook et al. (1974). The popular story of Surtsey has been told in a well-illustrated book by Sigurdur Thorarinsson (1967).

Figure 3.10 Discharge to ground from volcanic cloud

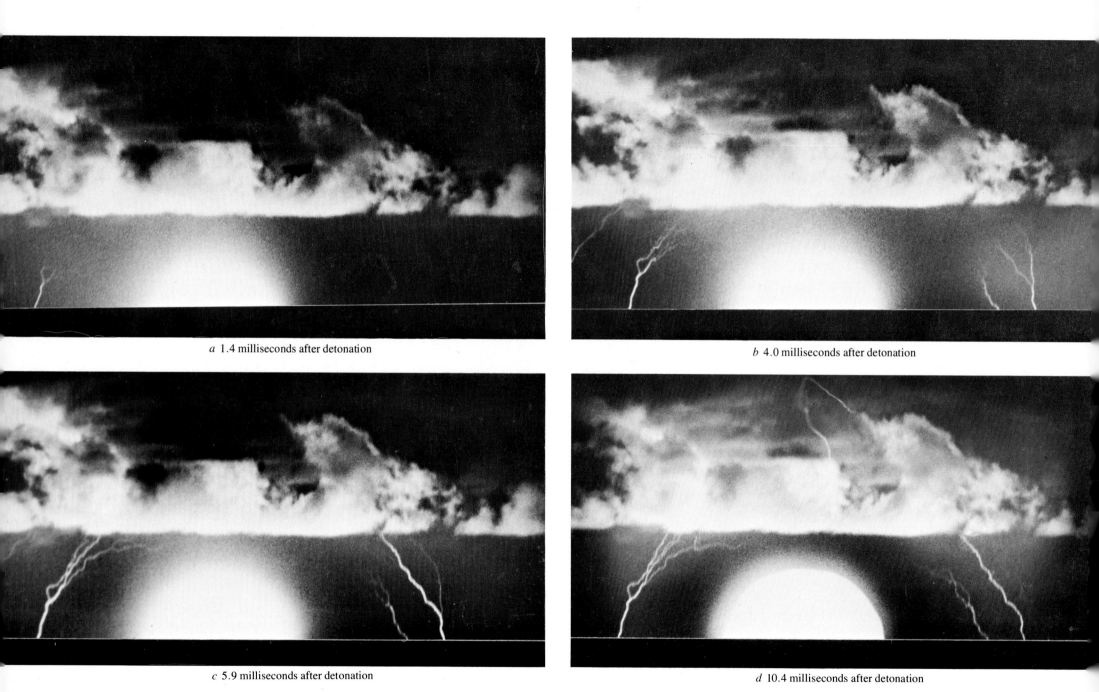

a 1.4 milliseconds after detonation

b 4.0 milliseconds after detonation

c 5.9 milliseconds after detonation

d 10.4 milliseconds after detonation

Figure 3.11 Lightning induced by thermonuclear fireball

Lightning induced by a thermonuclear detonation (Fig. 3.11). On 31 October 1952 the "Mike Shot" of Operation Ivy on Eniwetok Atoll produced an enormous fireball that triggered several flashes of lightning. Electrons liberated by a strong burst of gamma rays impinging on the air surrounding the fireball built up an electric field and charge in the atmosphere.

These photographs were taken from a distance of nearly 35 kilometers with a high-speed motion picture camera running at approximately 2200 frames per second. Each 16-mm frame was exposed for 100 microseconds with a 102-mm telephoto lens. The first frame shown here was exposed about 1.4 milliseconds after detonation; the second, third, and fourth frames here followed at intervals of 2.6, 1.9, and 4.5 milliseconds, respectively, as shown (D = time of detonation). Thus the series covers approximately 10 milliseconds in the development of the fireball and associated events. An approximate scale can be established on these reproductions by noting that the altitude of the cumulus cloud base was about 600 meters, and the bright flash to the left of the fireball was 900 meters from the point of detonation. (The image of the fireball is greatly overexposed in these prints.)

The discharges were of the upward streamer type, similar to those observed to flash from tall structures during normal thunderstorm conditions (see Fig. 2.15). The tips of instrumentation towers in the neighborhood of 1 kilometer from ground zero were apparently the elevated points that initiated these flashes. The tips of the streamers moved upward at approximately 1.5×10^5 meters per second; this velocity is comparable to that which exists for naturally occurring lightning of the same type.

Results of a careful study of the film and other data, together with an attempt to explain what was observed, have been published by Uman et al. (1972). The investigators were able to derive an electric field intensity of nearly the value usually regarded as necessary to initiate lightning — that is, 10^5 volts per meter. However, their estimate of the amount of charge available appeared to be insufficient to account for the number of discharges actually observed. A later analysis by Hill (1973) attempted to remove this discrepancy. *(Reproductions courtesy of Frank Berry and William Regan, Los Alamos Scientific Laboratory, from photographs obtained by Edgerton, Germeshausen, and Grier, Inc.)*

Multistroked flash recorded with a moving camera
Courtesy of New Mexico Institute of Mining and Technology

Figure 4.1 Flash to ground resolved into strokes

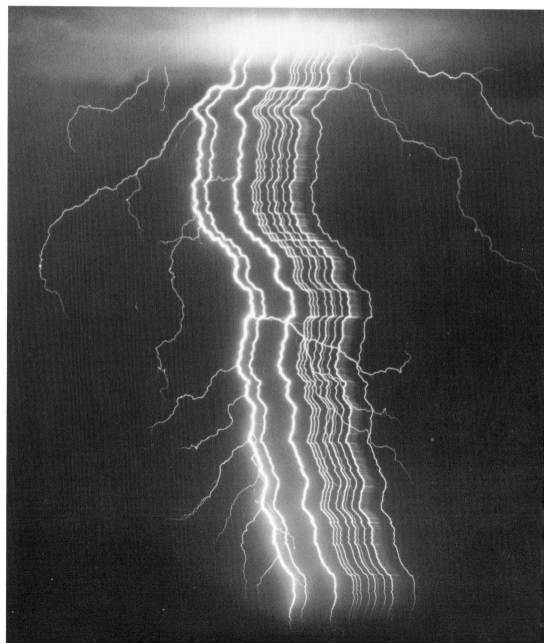

TIME-RESOLVED

AND OTHER SPECIALIZED PHOTOGRAPHS

4

Bright flash to ground resolved into strokes (Fig. 4.1). A camera was mounted to swing back and forth over a short arc while its shutter was held open. The resultant image motion streaked the image of the flash so that twelve component strokes are clearly separated over an interval of 0.6 second. Time increases from left to right in this display. The number of strokes and the interval over which they are spread are such that the eye might have perceived a flicker in the flash. This is a common

observation, intrinsic to a flash made up of many strokes, and is not due to some peculiarity of vision.

Note that branching occurs almost exclusively on the first stroke; subsequent strokes follow the same path, that of the first discharge. Note also that the image of the eleventh stroke is streaked out; this is *continuing luminosity* caused by the persistence of current flowing in the channel for approximately 60 milliseconds after the main pulse. The camera lens, an Aero

Figure 4.2 Lightning photographed with swing-camera technique

Tessar F/6, had a focal length of 61 centimeters; image motion was approximately 5 centimeters per second. *(Photo courtesy of New Mexico Institute of Mining and Technology, Socorro.)*

The highly detailed photograph in Fig. 4.1 was obtained with a large camera mounted on a heavy stand and made to oscillate by a mechanical linkage. However, the first pictures of lightning flashes streaked out into component strokes were obtained by simply pointing a small camera with shutter open toward an active thunderstorm, at the same time swinging it slowly back and forth until a flash appeared in the viewfinder, then closing the shutter. Results of this simple experiment were first published by L. Weber and H. Hoffert, German and English physicists, respectively — both in 1889. Evidently they had noticed the flickering of some lightning flashes and correctly inferred that this phenomenon is due to a quick succession of separate discharges that make up the total flash.

Photograph of lightning taken with a 35-mm camera, hand-held in back-and-forth motion (Fig. 4.2). Modern high-speed color film for 35-mm cameras makes it easy to get interesting and beautiful photographic effects with the open-shutter, swing-camera technique — given the circumstances of an active thunderstorm with flashes clearly visible below the base of the clouds. Point the camera in the direction lightning is flashing, open the shutter, and start moving the camera back and forth *slowly;* that is, sweep over an arc of about 20 degrees and take between 5 and 10 seconds to move in each direction. On a dark night, and in the absence of excessive foreground illumination, the shutter can be kept open for a minute or two before the film should be advanced to avoid fogging. Meanwhile, if a flash occurs in the viewfinder, *don't close the exposure instantly;* count to three. Flashes frequently last several tenths of a second; occasionally one will continue for a second or two. Too quick a closure of the shutter may cut off the last several strokes.

Figure 4.2, showing a four-stroke flash nicely resolved into components, was obtained in just this way. The camera was loaded with Ektachrome-X (ASA 64), the lens stop was set at f/11, the shutter was on "B" and was held open for about 15 seconds before the flash appeared in the viewfinder and the exposure was closed. *(Photo courtesy of William L. Taylor, National Severe Storms Laboratory, Norman, Oklahoma.)*

Ground flash, with cloud flashes above, photographed with a moving camera (Fig. 4.3). The ground flash in this illustration appears as a single stroke followed by a very brief (50 milliseconds or less) continuing luminosity — a relatively simple event. But above it, in and below the clouds, there appears a complex network of air discharges in which continuing current streamers precede bright events of short duration known as intracloud return strokes. (Time increases from left to right.)

This photograph, together with associated data on electric field changes, and a model for the process of intracloud discharge, have been analyzed in detail by Ogawa and Brook (1964). *(Photo courtesy of New Mexico Institute of Mining and Technology, Socorro.)*

Figure 4.3 Lightning flashes showing intracloud return strokes

(53)

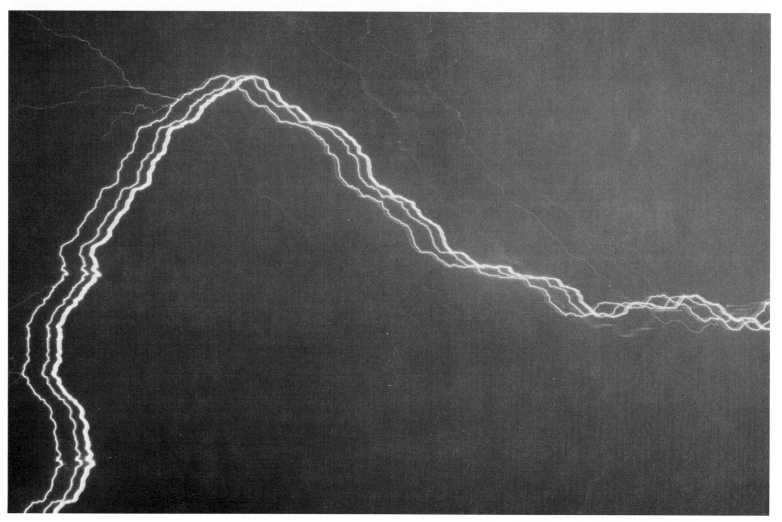

Figure 4.4 Five-stroked flash resolved in 0.25 second

Portion of a lightning flash, apparently emerging from the cloud base, and extending horizontally for some distance before it turns downward and out of the picture. (Fig. 4.4). Time increases from left to right in this photograph taken with a moving camera. Stroke 1 is branched, stroke 2 has very short (less than 20 milliseconds) continuing luminosity, and what may appear at first glance to be one final stroke is actually three component strokes joined by continuing luminosity. The total flash occurred within about 250 milliseconds. *(Photo courtesy of New Mexico Institute of Mining and Technology, Socorro.)*

Continuing luminosity following a flash (Fig. 4.5). Continuing current following the second stroke kept the channel luminous for approximately 380 milliseconds

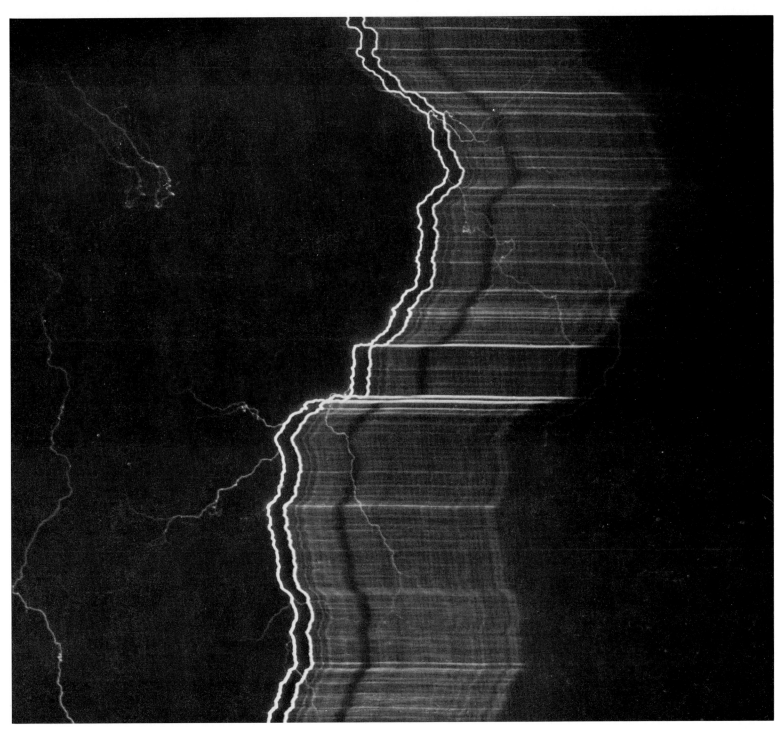

Figure 4.5 Continuing luminosity following flash

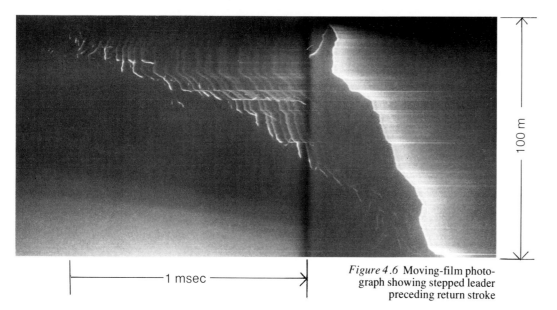

100 m

1 msec

Figure 4.6 Moving-film photograph showing stepped leader preceding return stroke

Figure 4.7 Fixed-film photograph of flash shown in Figure 4.6

(strokes separated by 25 milliseconds). Note that the light from the channel fluctuated irregularly and even ceased briefly as shown by the dark space in the moving camera record. The brightest horizontal streaks were probably produced by increased effective exposure where the channel is oriented along the direction of image-motion, or in the line of sight. It is possible that there are short lengths, or beads, in which the channel is intrinsically more luminous. However, that supposition cannot be established from this photograph alone. *(Photo courtesy of New Mexico Institute of Mining and Technology, Socorro.)*

Lightning flashes to ground, photographed with moving film to show the leader-return stroke sequence (Figs. 4.6-4.9). In Figure 4.6 a film belt, moving at the rate of 27 meters per second in the focal plane of a 50-mm camera lens, recorded the development of the faint stepped leader (on the left) that preceded the bright return stroke. In order to reproduce the extreme difference in brightness between the two events recorded on the original negative, the photographic printing process used a *dodging mask* to equalize the exposures on the paper. The diffuse vertical band was caused by the mask. Figure 4.7 is a fixed-film photograph of the same section of lightning channel, about 100 meters in length. The contact point with ground is hidden behind a parapet on the building from which this photograph was taken.

Figure 4.8 is a moving-film photograph of a downward negative leader to a tower, with several leaders ending in branches. The stepped leader associated with the brilliant return stroke is the one that joins the channel at point *J*. Its origin can be traced backward to the upper left corner of the picture, where it becomes lost among several other stepped leaders that produced no return stroke but ended in branches off the main channel. As in Figure 4.6, this reproduction is made from a print that was dodged to compensate for extreme overexposure of the return stroke. However, in Figure 4.8 the two leaders are bright enough to follow through the masked area: one to the point marked *J*, and another to one of the lower branches off the main channel. The last step of the principal negative leader is about 35 meters above the tip of the tower; it was joined by an upward connecting streamer, positively charged and much less luminous than the negative leader and hence not recorded on the film. This junction process actually initiates the return stroke. Figure 4.9 is a fixed-film photograph of the return stroke and its branching. *(Photos courtesy of Swiss High Voltage Research Committee, Zurich.)*

Many photographs similar to Figures 4.6 and 4.8 have shown that a downward-propagating lightning flash begins with a faintly glowing electrical discharge that proceeds in a step-like fashion, starting from somewhere in the charged volume at the base of a cloud, a region of predominantly negative charge in a typical thunderstorm situation. The average speed of the steps is approximately 1.5×10^5 meters per second. Each step travels about 50 meters, with a pause on the order of 50 microseconds between each one. The tip of the discharge channel brightens at the onset of each step, only to fade out with the rest of the leader before the whole process is repeated in the next step.

A distance of 3 to 6 kilometers typically occurs between the ground and the charged region in the base of a cloud, and it therefore takes the stepped leader about 20 to 40 milliseconds to come within striking distance of the ground, or some object standing above it — a

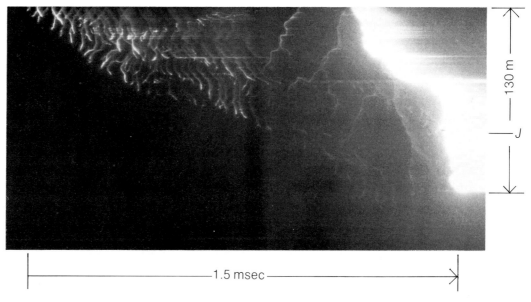

Figure 4.8 Moving-film photograph showing stepped leader-return stroke sequence

Figure 4.9 Fixed-film photograph of flash shown in Figure 4.8

building, a tree, a metal tower, etc. At that moment, the intense electrical field between the ground and the tip of the leader breaks down the air in a great spark-over. This initiates the return stroke, whose luminosity propagates upward into the cloud base, along the tortuous conducting path previously made by the leader process. The return stroke moves upward as a wavefront of ionization and radiation (light), with a speed that is typically one-tenth to one-third that of light. Thus, while the comparatively sluggish stepped leader process may have taken 20 milliseconds to travel from cloud to ground, the rapid motion of the return stroke covered the same distance in about 70 *microseconds*! (In looking at these time-streaked photographs, it is important to keep in mind that the leader and the return stroke follow the same path; it is the rapid horizontal motion of the film that separates the image of one from the other.) Leaders that fail to get close enough to the ground to initiate the principal discharge, or return stroke, end in branches, or air discharges.

Figure 4.10 Dart leader preceding a return stroke

Dart leader following the channel established by one stroke and triggering another (Fig. 4.10). If fewer than 100 milliseconds elapse after the passage of current in a return stroke process, a dart leader may move rather smoothly down the path of residual ionization. The speed of a dart leader is about 2×10^6 meters per second, or an order of magnitude greater than the average speed of a stepped leader. In the example shown in Figure 4.10, recorded with a 139-mm telephoto lens and a film-loop moving 25 meters per second, the dart leader traveled approximately 2500 meters from cloud-base to ground in 1.25 milliseconds (compare with the stepped leader in Figure 4.6 that required 1 millisecond to travel less than 100 meters). *(Photo by Clyde N. Richards, courtesy of Institute of Atmospheric Physics, University of Arizona, Tucson.)*

A close examination of the lowest third of the leader reveals that, instead of maintaining a continuous path, it began to exhibit short discontinuities, or steps. This is typical of what occurs as the dart moves downward along the path established by the preceding stroke and encounters regions of lower ionization, where there has been appreciable time for free electrons to recombine with ions. As the dart meets higher and higher resistance to current flow, it begins to act like a stepped leader. If more than a tenth of a second has elapsed since the previous return stroke moved up that part of the channel, the leader has to revert entirely to the process illustrated in Figures 4.6 and 4.8, thereupon establishing a new channel for the next return stroke. This stroke will surge upward from a different point on the ground but will rejoin the previous channel just where the dart changed over into a stepped leader. Thus, an ordinary photograph will show a forked lightning channel as in Figs. 2.2 and 2.4).

Twenty-six strokes in one lightning flash that lasted nearly two seconds (Figs. 4.11, 4.12). This remarkable event was recorded on the evening of 26 July 1959, at Socorro, New Mexico, with two cameras having different scales of time-resolution. Figure 4.11 is similar to Figure 4.1; the moving camera had a focal length of 61 centimeters and image motion of 5 centimeters per second. Figure 4.12 is made up of five sectors of a rotating film exposed through a lens of 75 millimeters focal length. The 9″ x 9″ flat film holder turned at 2 rotations per second in the focal plane so that the images are spread out on a circle with a 10-centimeter radius. The resultant film transport speed along the image arc was approximately 30 centimeters per second. Both figures are reproduced to original scale, from contact prints.

In these reproductions, time increases from left to right. The first stroke is branched, as usual, but it is not the brightest. When a flash consists of several strokes, the first is usually the brightest; here, the last stroke has that distinction. The original negative shows dart leaders for all strokes subsequent to the first; the leader associated with the sixth stroke is especially well shown even in this reproduction (Fig. 4.12).

This complex event has been investigated in detail and reported upon by Workman et al. (1960). The study concluded that the flash involved at least four, and possibly six, times the quantity of charge normally associated with a single flash, or on the order of 100 to 200 coulombs of negative charge transferred to ground by these twenty-six strokes. Cloud electrification was probably tapped from several adjacent cells in quick succession, thus producing a closely connected series of discharge events. A later and more extensive study of this type of photograph by the same investigators (Workman et al., 1962*a, b*) gave more information on

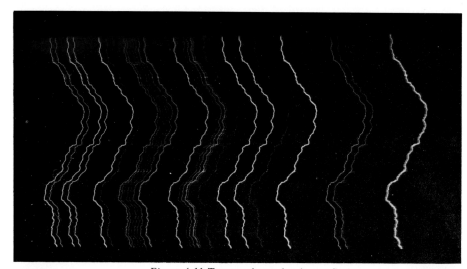

Figure 4.11 Twenty-six strokes in one flash

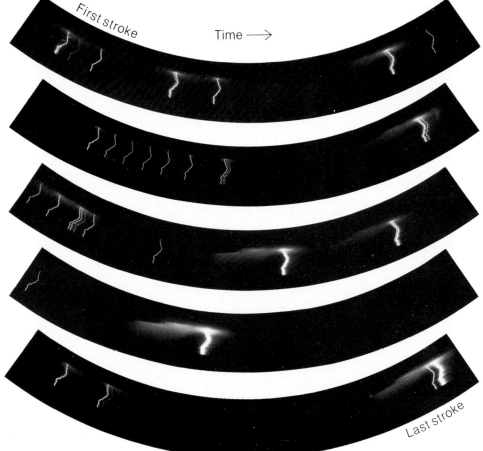

Ground

Figure 4.12 Twenty-six-stroked flash streaked out to show dart leaders

the behavior of continuing luminosity and associated currents. *(Photos courtesy of New Mexico Institute of Mining and Technology, Socorro.)*

Upward propagating negatively charged leader and its following stroke, originating from the tip of a metal tower (Figs. 4.13-4.15). A leader carrying negative charge, either downward from a cloud or — as in this case — upward from a metal tower, exhibits a characteristic stepping process in its path through the air, as shown in Figure 4.13. This photograph was taken with the same type of moving-film camera that produced Figures 4.6 and 4.8 and similarly shows the distinct steps by which the discharge channel develops. Note that the channel luminosity between steps is negligible. The tower tip is 70 meters above the ground and 400 meters from the camera. Upward strokes following the path of a negative, stepped leader exhibit branching — just as do the negative downward flashes (see Fig. 2.15 for a good example of an upward branching flash from a tower). Figure 4.14 is an enlargement showing the early phases in the development of the same leader as shown in Figure 4.13, with faint corona discharges, which look like tufts of hair, at the tip of each step. Figure 4.15 is a fixed-film photograph of the same upward propagating flash. *(Photos courtesy of Swiss High Voltage Research Committee, Zurich.)*

Upward propagating, positively charged leaders and their following strokes originating from the tips of metal towers (Figs. 4.16, 4.17). In Figure 4.16, a moving-film photograph, there are two leaders — one from each of two towers in the field of view. Leaders

100 m

Figure 4.13 Moving-film photograph of upward propagating negative leader

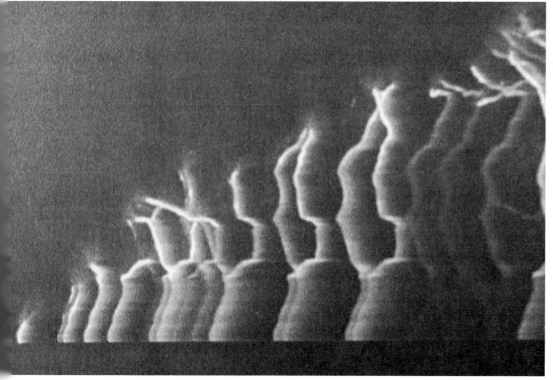

Figure 4.14 Early phases in development of upward propagating negative leader

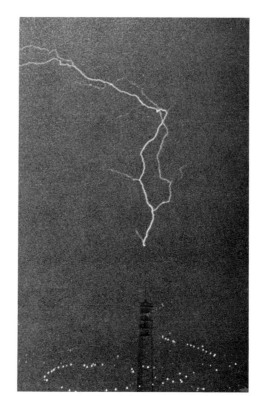

Figure 4.15 Fixed-film photograph of flash shown in Figures 4.13 and 4.14

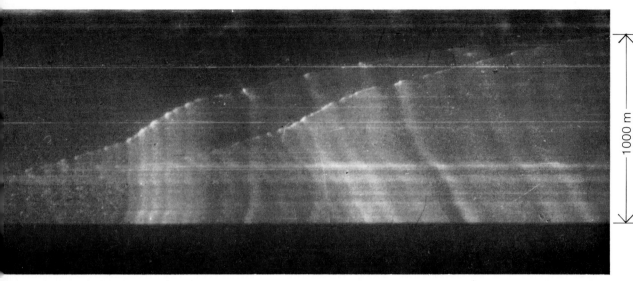

Figure 4.16 Moving-film photograph of upward propagating positive leaders

— 1 msec —

1000 m

carrying positive charge do not move in the pronounced stepped manner of their negatively charged counterparts. Positive leaders give out considerably less light than negative ones and show a comparatively smooth, continuous development with less variation in overall brightness as they progress upward. Compare this view with Figures 4.6, 4.8, and 4.13.

Figure 4.17 is a fixed-film photograph of the strokes that followed the leaders recorded by the streak camera. This type of stroke tends to exhibit less branching than do downward negative strokes. When the positive leader contacts a negatively charged region at the base of a cloud, the current rises to a fairly steady value of a few hundred amperes, often lasting for several tenths of a second. The amount of charge transferred in such a positive discharge can be very large, as much as 100 to 300 coulombs. In contrast, in the case of a typical return stroke following a negative stepped leader, currents usually peak at tens of thousands of amperes, last for only a few tens of microseconds, and transfer perhaps 10 to 20 coulombs of charge. *(Photos courtesy of Swiss High Voltage Research Committee, Zurich.)*

A comprehensive photographic study of leaders and their return strokes correlated with current measurements has been carried out by Karl Berger and his associates in the Swiss High Voltage Research Committee for lightning flashes to metal towers atop Mount

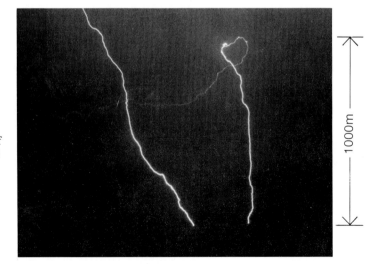

Figure 4.17 Fixed-film photograph of flash shown in Figure 4.16

1000m

San Salvatore in southern Switzerland. Descriptions of this work (in English) have been published by Berger (1967) and Berger and Vogelsanger (1969). Original papers (in German) were published by Berger (1966, 1972, 1973). These investigations have been summarized and reviewed in detail by Uman (1969*a*, Chap. 2).

Daylight photography of lightning (Figs. 4.18, 4.19). Figure 4.18 shows a lightning flash photographed through an H-alpha filter during a daytime thunderstorm. The optical spectrum of lightning reveals that one of the strongest features in visible wavelengths is the hydrogen emission at 6563 Å (angstroms) — the hydrogen alpha, or first member of the Balmer series (see, for example, Fig. 6.7). Dry air contains less than .0001 percent hydrogen, by volume, but when a heavy spark passes through moist air, as in and around a thunderstorm, a significant amount of this gas is produced for a very short time by the dissociation of water vapor into two parts atomic hydrogen and one part oxygen. It is these free atoms of hydrogen that are excited to radiate by the strong electric current passing through the lightning channel. This excitation process gives rise to the strong H-alpha as well as other spectral emissions characteristic of hydrogen, oxygen, nitrogen, argon, all of which are present in the atmosphere.

Figure 4.18 Daylight photograph of lightning through H-alpha filter

Figure 4.19 Daylight
photograph of lightning without filter

The photograph was taken with a narrow passband filter in front of the lens. The filter transmitted about 16 Å on each side of the strong emission line of hydrogen, which appears to the eye as a very deep red. The film was Kodak SO 375, an emulsion with low-speed, high-contrast properties and relatively high spectral sensitivity in the region of 6500-6600 Å. This combination increased the exposure of the lightning channel relative to the daylight background below the cloud and, with the lens stopped down to F/11 or less, the shutter could be kept open about half a minute without overexposing the film to an unacceptable level. In this particular case, the shutter was open only a few seconds before a flash appeared in the viewfinder and the photographer closed the shutter. This photographic method works well in the vicinity of active thunderstorm cells where lightning flashes tend to appear in approximately the same direction, as is the case in southern Arizona where this photograph was taken. *(Photo courtesy of George Marcek, Tucson.)*

Figure 4.19 is a daytime photograph of lightning taken without an H-alpha filter or any other special device. Under a very dark, active thundercloud, with a slow, contrasty film and lens stopped down to F/16 or less, the photographer can keep the shutter open for several seconds before the film becomes badly overexposed. In a few cases out of many attempts, a light-ning flash will be captured as shown here. The film used was Adox KB-14. Note the blurred images of the two flags as they fluttered during the time the shutter was open. *(Photo courtesy of George Marcek, Tucson.)*

In rare instances it may happen that lightning will strike somewhere in the field of view of a camera when its shutter is snapped — this is sheer luck. But it is possible for the photographer to do much better than chance alone can afford. In the first place, when an active thunderstorm develops, several flashes may come out of the same cloud within a minute or two and appear in very nearly the same direction. Secondly, it is very common for a flash to be made up of several strokes; these may occur over an interval of several tenths of a second (see Figs. 4.1 and 4.4). The photographer's strategy is to point his camera toward an active thunderstorm, to wait, and then to snap the shutter the instant he sees a flash in the viewfinder. The shutter speed should be about one-half of a second, with the lens stop set to match the ambient illumination and film speed. Reaction time — eye-brain-finger-shutter — will not be fast enough to catch the first stroke, but, in the case of a multistroked flash, the chances are very good that one or more of the subsequent strokes will occur during the time the shutter is open. Techniques for daylight photography of lightning have been published by Salanave and Brook (1965) and Krider (1966).

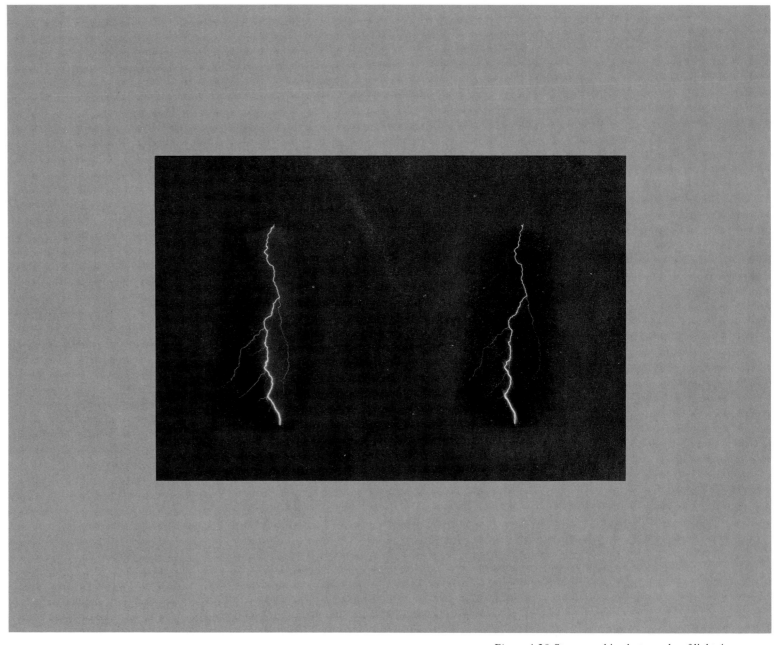

Figure 4.20 Stereographic photographs of lightning

Stereographic photographs of lightning (Fig. 4.20). Two cameras, separated by approximately 300 meters and pointing toward the same thunderstorm, recorded the same flash about 10 kilometers away. Both cameras had a focal length of 310 millimeters; the photographs have been reproduced on slightly different scales in order to match them for stereoscopic viewing. The apparent vertical length of the main lightning channel is about 2500 meters.

The photographs can be viewed best by using what is known as a pocket stereoscope. This instrument is designed to accommodate a typical interpupillar distance of 65 millimeters; it is widely used in college-level courses in geology and geography, and can usually be purchased inexpensively wherever surveying and engineering supplies are sold. Stereoscopic viewing can also be accomplished without optical aid as follows: hold the pictures at a comfortable reading distance (i.e., 30 to 50 centimeters away) and try to fuse the two views by looking at them as though they were at infinity — that is, by relaxing binocular vision convergence while at the same time focusing on the page. It is easier said than done but makes an interesting experiment. Persons with monocular vision cannot do this, and problems with ocular muscle control make it difficult to fuse the two images. Persons who are myopic (nearsighted) and wear corrective lenses can achieve an advantage by removing their glasses and holding the page close enough to get a sharp focus — if their eyes have equal power, no astigmatism, and good muscular balance.

Stereographs such as these enable the viewer to appreciate the tortuosity of lightning, the twisting and bending path that follows the ionization established by the stepped leader. Figures 5.1–5.6 show channel tortuosity in more detail. *(Photos courtesy of New Mexico Institute of Mining and Technology, Socorro.)*

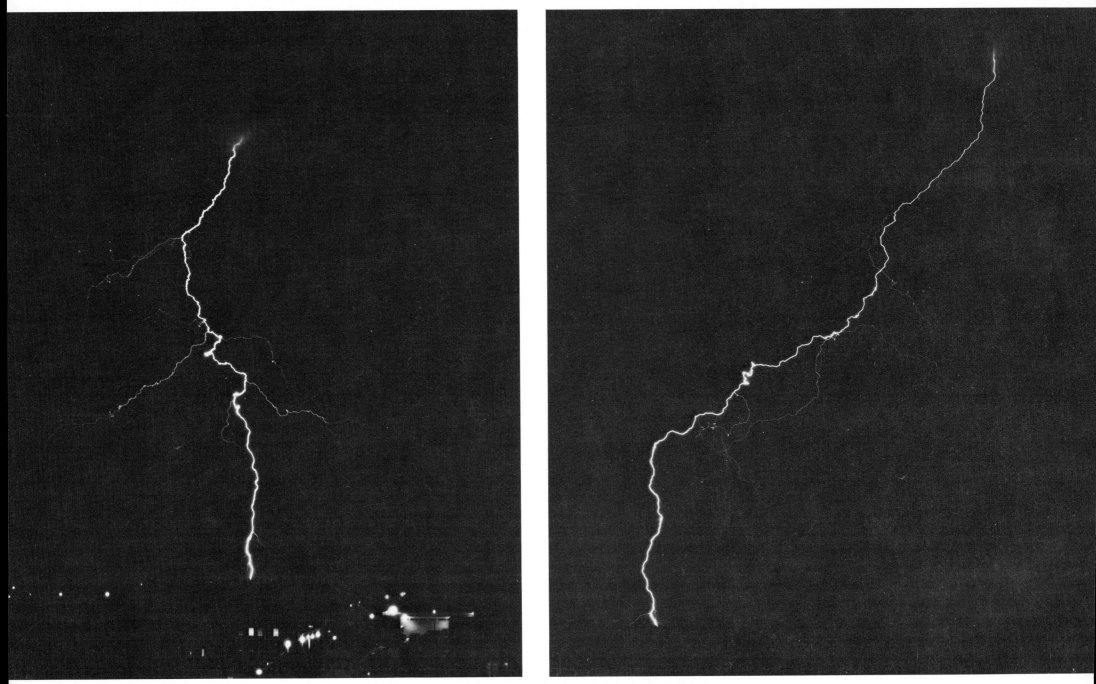

Figure 4.21 Crossed-view photographs of a typical cloud-to-ground flash

Figure 4.22 Crossed-view photographs of a nearly vertical flash

Paired views of lightning flashes, photographed with identical cameras but from different directions (Figs. 4.21-4.23). In these pairs, the view on the left was obtained with a camera (designated No. 1) whose axis was aligned 29.8 degrees in azimuth clockwise around the horizon from the pointing of the other camera (No. 2). The distance of the flashes from both cameras was 4 to 5 kilometers in all cases, so that the resolution of detail along the discharge channels is about the same; the scale of reproduction is on the order of 300 meters per centimeter. Only the direction of viewing is notably different, with one line of sight making an angle of nearly 30 degrees with the other. Both cameras had a focal length of 155 millimeters.

Figure 4.23a Air discharge

The pair of photographs in Figure 4.21 shows a typical cloud-to-ground flash. On the left the channel appears to be nearly vertical, but a glance at the picture on the right (from camera No. 2) makes it clear that the channel viewed from camera No. 1 must be considerably foreshortened. Thus the lowest quarter of its length is closer to the observer than the upper three quarters.

The pair of photographs in Figure 4.22 shows a flash whose channel is comparatively straight and vertical. Notice the relatively small differences in overall shape of the channel as represented in the left and right views. In these photographs the scale of tortuosity is on the order of several dekameters; this is generally regarded as the *mesoscale* in discussions of the

Figure 4.23b Same air discharge
viewed from another direction

characteristic path of a lightning discharge. For resolution of detail that is one or two orders of magnitude smaller (the *microscale*), see Figure 5.6.

The pair of pictures in Figure 4.23 shows a fine example of an air discharge. A comparison of the two views gives a clear idea of the relative positions and alignments in the structure of this intricate lightning channel. Note especially that the tight knot seen in Figure 4.23*b* can be unraveled by referring to the companion view, Figure 4.23*a*. *(Photos courtesy of Institute of Atmospheric Physics, University of Arizona, Tucson.)*

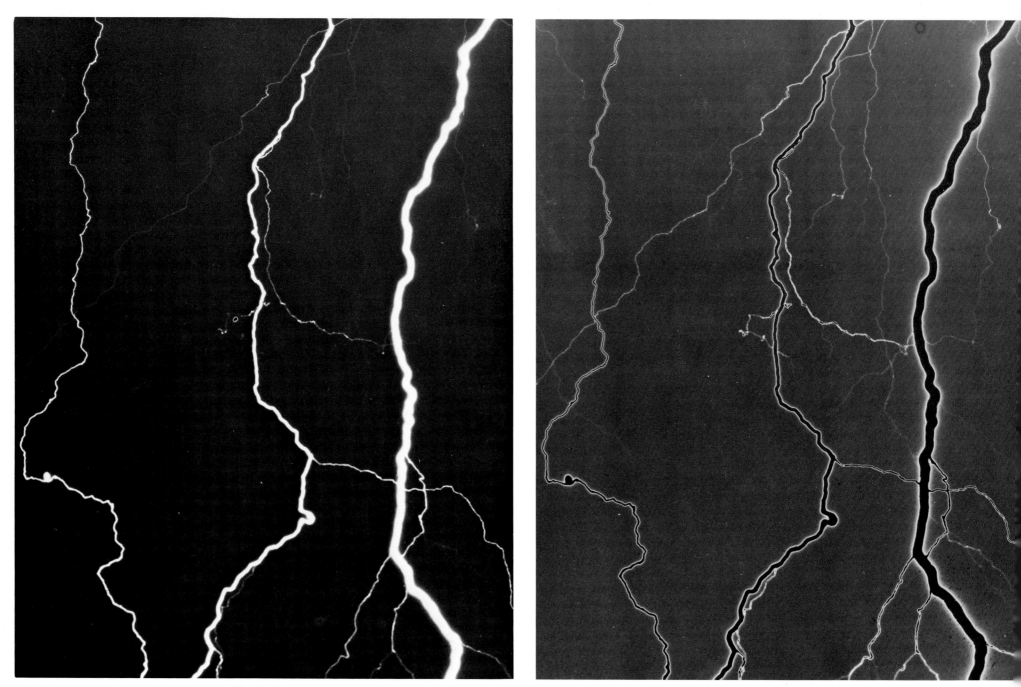

Fine structure and tortuosity in lightning channels as shown in normal print (*left*) and print with Sabattier reversal effect (*right*)

Prints by C. S. Rainwater, Solarol Corporation, from University of Arizona photograph

THE STRUCTURE OF LIGHTNING

TORTUOSITY OF CHANNELS

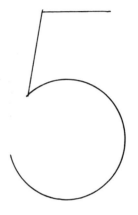

Detailed structure of lightning discharges to ground, as shown by photographs taken with a large telephoto camera (Figs. 5.1-5.5). These five photographs were made during the summers of 1966-1969 with an aerial camera equipped with an F/6.3 lens of 1220-mm focal length and situated at the University of Arizona, Tucson. At the time this was probably the largest camera, with the highest angular resolution, ever used for the direct photography of lightning. The photographs show the intricate structure, or tortuosity, that is characteristic of the lightning channel. The camera was pointed slightly upward, at an angle of 3.5 degrees above the horizon. The center of each photograph corresponds to an elevation above the horizon of approximately 1/16 of the distance to the flash. In most cases, the distance to the lightning channel (assumed to be vertical between the ground and 1 or 2 kilometers up) is only roughly known, with errors of 30 percent being quite probable.

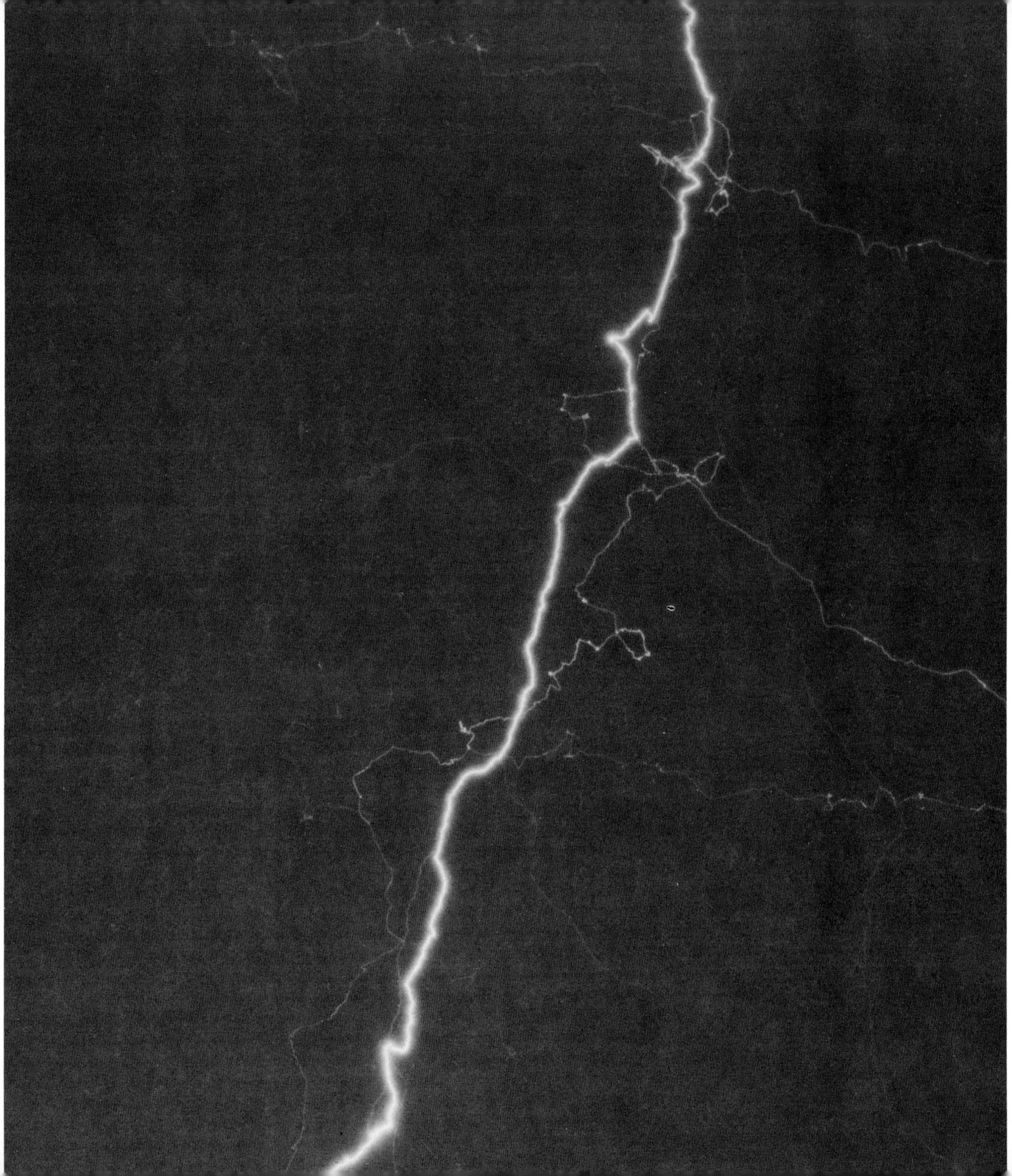

Figure 5.1 Distant flash photographed with 1220-mm telephoto camera

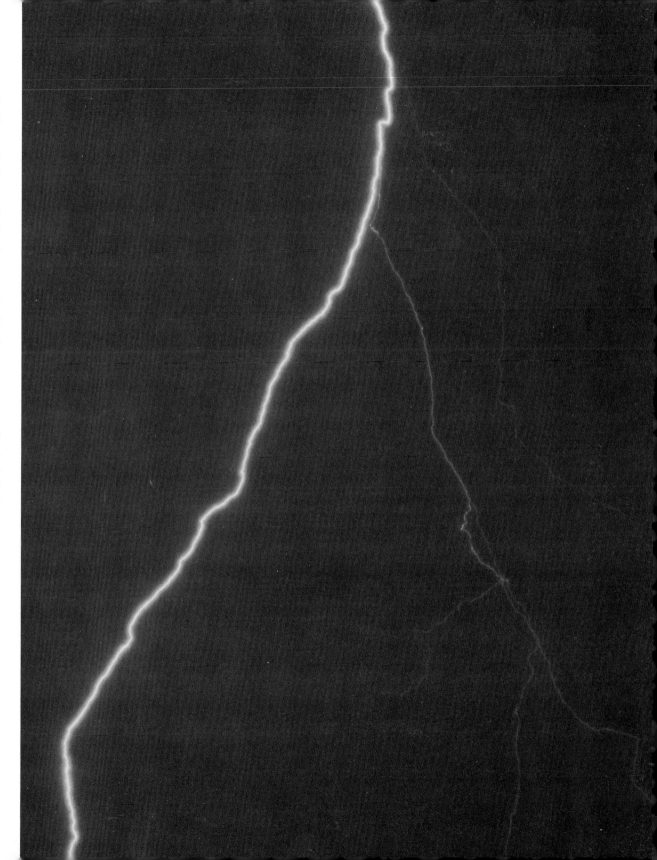

One can, however, take these photographs as representative of the appearance of a lightning channel about 300 meters above ground for the nearer flashes, and 1000 meters or more above ground for the more distant ones. As the land surface around the lightning observatory is flat or only slightly hilly out to nearly 15 kilometers in all directions, it is clear that none of these views shows the lightning near its contact with the ground (for an example of a ground contact photographed with this camera, see Fig. 7.7).

Higher spatial resolution can be achieved with smaller cameras, but only in cases of chance strikes by very close lightning in the field of view, or by focusing the camera on the tip of a nearby tower or other elevated point (see Figs. 3.6-3.8, 4.6-4.9, 4.13-4.15, 5.6, 5.7, 7.1, 7.2, and 7.9).

The photograph in Figure 5.1 was made on Kodak Plus-X film, with a yellow filter and the lens at full aperture, i.e., F/6.3. It was processed in a phenidone developer, which has the characteristic of producing a low-contrast image (low gamma) without sacrificing the fainter features of the subject. That is, the latitude of the film is greatly extended without a significant loss of speed. This property is particularly well suited for recording the great range in luminosity of details associated with lightning: the principal channel with sharp bends and bright spots along its course, the very fine branches and their complex structure, and any gradations representing cross-sectional variations in the image of the channel. In this photograph the channel appears "soft," that is, with an apparent outward decrease in brightness (see also Fig. 5.2, made in the same way but several nights later). Whether this halo is due to an enveloping corona discharge, which is questionable, or

Figure 5.2 Wide range of channel details resolved down to one-meter dimension, recorded with fine-grain film and low-contrast developer

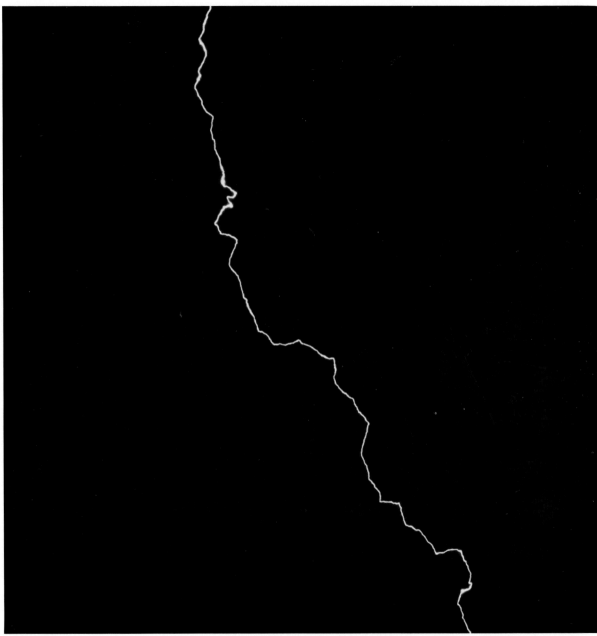

Figure 5.3 Detail similar to that in Figure 5.2, but recorded with high-contrast developer and filter to emphasize H-alpha emission from the lightning channel

to the diffusion of light through mist and rain in the line of sight, which is more probable, the fact is that the same film processed in an ordinary developer to a normal or high contrast would not have shown this detail as clearly, if at all. (Halation, i.e., light scattering in the emulsion or its backing, would have a somewhat different appearance and could be recognized as such.)

The distance to this flash was estimated to be about 15 kilometers. The scale of reproduction is 10 millimeters to approximately 35 meters at the lightning. Some of the finer structural features, such as sharp kinks and bright spots, can be resolved down to dimensions of 2 or 3 meters.

Figures 5.2 and 5.3 provide a clear comparison between the results of photographing lightning with different combinations of film, filter, and developer, while using the same camera at nearly the same distance from the flashes. In both cases, the flashes were about 4 kilometers from the camera. The scale on these reproductions is 10 millimeters to approximately 10 meters at the flash; details in the structure of the channel separated by about 1 meter can be made out. Lens stops were f/8 and f/11, respectively, and are indistinguishable as far as image-quality is concerned.

The significant differences are in the film types, the filters, and the developers used. Figure 5.2 was exposed on Plus-X film, with a light yellow (No. 12) filter and processed in a phenidone developer. Figure 5.3 was exposed on a special film, Kodak SO-392, characterized by very fine grain and high contrast. Its spectral sensitivity peaks near H-alpha, at 6563 Å; the emulsion is especially useful in recording solar chromospheric phenomena such as prominences and flares. Then, for the purpose of lightning photography, a very deep red filter (No. 92) was used to limit the spectral range to a band from 6300 to 6800 Å. This has two significant effects: first, radiation from hydrogen atoms (i.e., from H-alpha) is emphasized in the exposure (see Fig. 6.7). Second, the range of wavelengths is so small that any residual chromatic aberration in the lens, however well corrected, becomes entirely insignificant, and the image can be focused very sharply. Finally, the film was processed in Kodak D-19, a developer that gives high contrast and maximum acuity in the photographic image.

Figure 5.4 shows image contrast between the extremes represented by Figures 5.2 and 5.3. Recorded on Plus-X film, with a red (No. 25) filter, and developed in Rodinal, the contrast in the image is moderate and preserves some gradation in the profile of the channel. The observing record for the night this picture was taken records only that the storm was "several miles away"; the resolution of details along the channel is probably in the range of 2 to 3 meters. The reader may try to decide whether the channel coming into the picture at the upper right crosses the other or osculates it and turns back toward the right.

Figure 5.5: What happened here? The original negative, obtained and processed with the same photographic parameters as for Figure 5.3, captured this little detail in a branch associated with a strong lightning discharge. There is no record of the distance from the camera, but it was probably about 15 kilometers. For that distance the scale on the reproduction is 10 millimeters to 30 meters at the lightning. A comparison with Figure 4.23 suggests that this image may be the end-on view of the nearly horizontal section of a very tortuous lightning channel. *(Photos courtesy of Institute of Atmospheric Physics, University of Arizona, Tucson.)*

Figure 5.4 Channel details recorded with a film and developer selected to give medium acuity and contrast

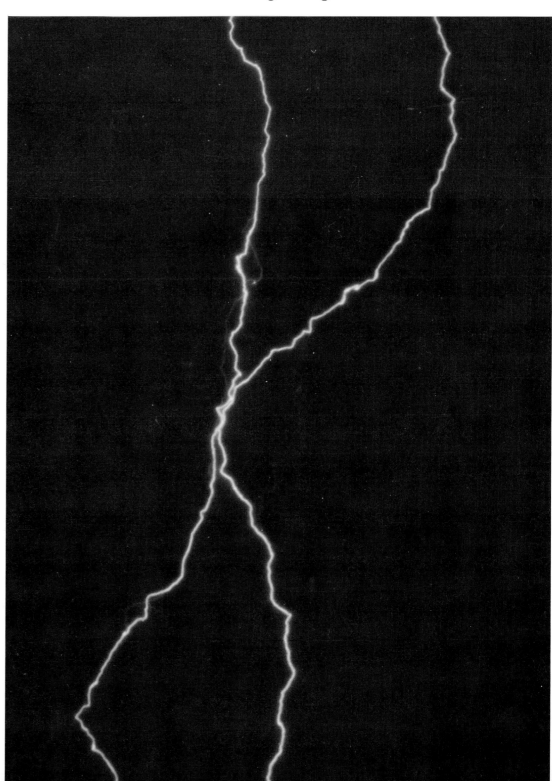

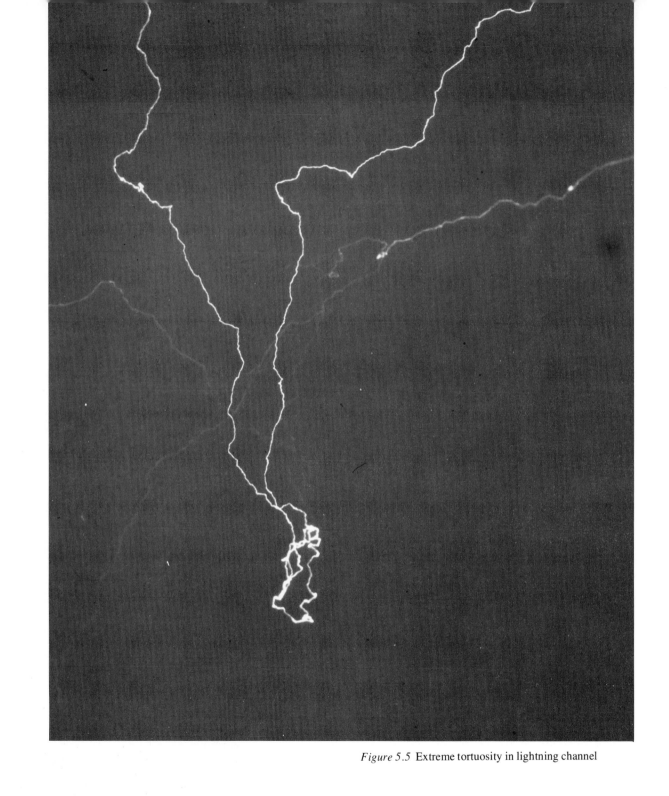

Figure 5.5 Extreme tortuosity in lightning channel

The microscale structure of tortuosity in a lightning channel (Fig. 5.6). This photograph was obtained in September 1961 by W. H. Evans and R. L. Walker, using a high-speed framing camera located on Mount Bigelow near Tucson, Arizona. It is the second exposure in a sequence of eight, each with a duration of 2 microseconds and with an interval of 77 microseconds between frames. Other details of the instrumentation are the same as those given for Figure 3.8.

The scale of this reproduction is 10 millimeters to 85 centimeters at the flash. Sinuous structures in the channel, with dimensions of 10 centimeters or more, can be readily made out. This is on the scale of micro-tortuosity. Evans and Walker made measurements across the channel's image with a densitometer, compared these results with those obtained from a simulated lightning channel of known width photographed with the same optical system and film, and obtained a value of nearly 12 centimeters for the average diameter of this flash. High-speed photographs of two other flashes to the Mount Bigelow tower, similarly calibrated and measured, gave diameters of 3 and 8 centimeters, respectively. See Evans and Walker (1963). *(Photo courtesy of Institute of Atmospheric Physics, University of Arizona, Tucson.)*

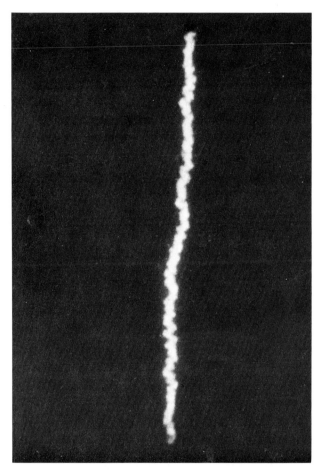

Figure 5.6 Microscale structure of tortuosity

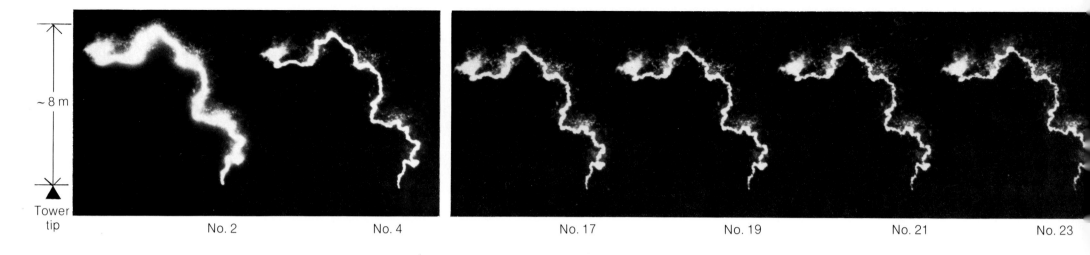

~8 m

Tower tip

No. 2   No. 4   No. 17   No. 19   No. 21   No. 23

Figure 5.7 Channel diameter and luminosity changes in a lightning stroke

Channel diameter and overall luminosity changes in a lightning stroke to a metal tower, as recorded by a high-speed framing camera (Fig. 5.7). This sequence of six photographs is from a remarkable series of 224 frames obtained for a stroke that lasted at least 18 milliseconds. The event was recorded on Mount Bigelow, Arizona, in August 1964 by the same high-speed framing camera and experimental arrangement described for Figure 3.8. The framing interval is 77 microseconds, so a pair of consecutive even- or odd-numbered frames spans 0.154 milliseconds. The six numbered frames shown (i.e., 2, 4, and 17, 19, 21 and 23) represent a total elapsed time of 1.62 milliseconds. Frame 1 was grossly overexposed.

Nearly a decade elapsed before the great amount of information in the original strip of film could be effectively extracted. As a practical matter, data reduction awaited the development of *scanning densitometers,* which can go over a photographic image quickly and display selected intervals, or steps, of density as a series of colors. This process results in a set of contours that can be interpreted, after calibration, in terms of luminosity levels over the light source as functions of position (*x-y* coordinates perpendicular to line of sight) and time (for successive images in a framing camera sequence). Such devices are called pseudocolor densitometers. See Figure 5.8 for an example of the resulting scan. *(Photo by W. H. Evans, courtesy of Institute of Atmospheric Physics, University of Arizona, Tucson.)*

Pseudocolor representation of a lightning photograph in five steps of density, from 0.0-0.5 (black) to 1.2 and over (blue) (Fig. 5.8). Photographic density, D, is defined as $\log_{10} 1/T$, where T is the fraction of incident light transmitted by the corresponding spot on the film whose density is being measured. For example, if $T = 10$ percent or 0.1, then $D = 1.0$. The scan was made on frame No. 3 in the series and, therefore, represents the state of the lightning channel midway between frames 2 and 4 in Figure 5.7.

The enhanced white line between the areas of red and brown is the isodensity contour at the value 0.7; it gives a first quantitative approximation to variations in apparent width of the channel. Notice, for example, the constriction at the lower end where the flash made contact with the metal rod at the top of the tower. This is typical of lightning channels at or near metal (a good conductor) as compared to their normal width in the atmosphere (a relatively poor conductor).

A calibration procedure to convert photographic densities to relative levels of intensity (light output) was required before the channel's true diameter profile and luminosity changes as a function of time could be determined. This work and its results have been described in detail by Orville et al. (1974). Briefly, their conclusions were that this lightning channel had a time-averaged diameter of 6.5 centimeters (measured at the position marked on the isodensity contour), and this diameter did not change significantly in the time covered by their measurements, i.e., in 1.6 milliseconds. On the other hand, image density and, therefore, channel luminosity dropped rapidly in the first 0.2 millisecond and then decreased gradually for the remainder of the measured interval. *(Reproduction courtesy of R. E. Orville, State University of New York at Albany.)*

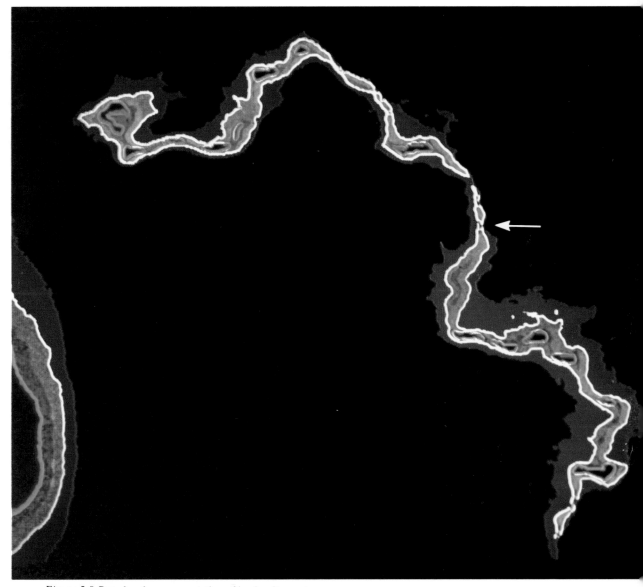

Figure 5.8 Pseudocolor representation of isodensity contours

The visible spectrum of lightning, with undispersed image of flash on the left

Courtesy of William Bickel, University of Arizona Department of Physics

THE PHOTOGRAPHIC SPECTRUM

6

Two historically significant spectra of lightning (Figs. 6.1, 6.2). Although visual observations of lightning spectra were made as early as 1868 and the first attempt at photography was in 1894, the observations were difficult to record and interpret until after the turn of the century. Then almost a half-century passed before anyone began to extract information from the spectrum about lightning's temperature, ionization states, and other physical parameters.

Considering the role that spectroscopy has long played in the understanding of the properties of luminous sources — especially inaccessible ones such as those astronomers deal with — it may seem surprising that lightning was so neglected as an object for spectrum analysis. However, much of the lag can be attributed to the sheer difficulty of observing the capricious "flight of thunderbolts" with the somewhat unwieldy and inefficient spectrographs that were typical of those available

Figure 6.1 Slitless spectrum of lightning obtained in 1901 with large objective prism

until around 1950. Then too, quantitative spectrum analysis had to wait for an understanding of the origin of atomic spectral lines, based on the development of quantum mechanics, which began in the mid-1920s.

The slitless spectrum of lightning shown in Figure 6.1 was obtained in 1901 at the Harvard College Observatory with the 20-centimeter (8-inch Draper) telescope fitted with an objective prism. The isochromatic emulsions of that time were not sensitive to red, or even orange, light; the photographic spectrum ended in the yellow region. On the other hand, the film was sensitive to ultraviolet light, but thick lenses and prisms strongly absorb those wavelengths that are already near the limit of visual perception. Thus the spectrum shown here extends only between 3900 Å (on the left) and 5600 Å (on the right). This spectrum was produced by direct photography of the lightning channel through a prism arranged so that its direction of dispersion was parallel to the horizon. In this way the thin source performed the function of a slit; that is, it provided for the formation of spectrum lines. (*Photo courtesy of Harvard College Observatory.)*

Lightning is a polychromatic source of radiation resembling a gaseous nebula more than the sun, and is a line-like source similar to a meteor trail. Thus the astronomical technique of *slitless spectroscopy* is admirably suited to the analysis of lightning. The "prismatic camera," or slitless spectrograph, gives not only a record of the wavelengths (lines) emitted by the source, but also gives information on how these emissions are distributed in the source. The slitless spectrum is therefore a two-dimensional display of wavelength versus position along the source. Of great practical importance is the fact that the slitless spectrograph is

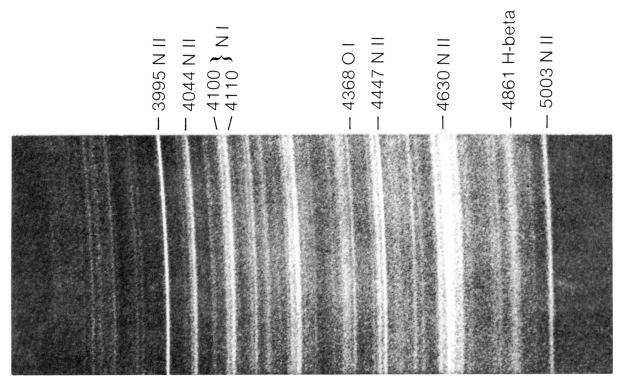

Figure 6.2 Spectrum of lightning photographed with slit spectrograph in 1917

not as wasteful of light as the conventional slit instrument, which must be used where the source is a surface of appreciable apparent width, such as a furnace, an aurora, or the sun. But the slitless instrument has the disadvantage of not providing a comparison, or standard, spectrum for the precise determination of wavelengths and the straightforward identification of chemical elements in the source (see Fig. 6.2 for a slit spectrum of lightning with emission lines identified). Figure 6.2 is the first photograph of the spectrum of lightning made with a slit spectrograph, obtained in 1917 by V. M. Slipher at the Lowell Observatory in Flagstaff, Arizona. The instrument, without the usual telescope in front of the slit, was pointed in the direction of a distant thunderstorm. Flashes of light diffused into the slit and collimator over an interval of several minutes and exposed a spectrum on the photographic plate. The result shown here is therefore a composite of many individual spectra.

The reproduction in Figure 6.2 is an enlargement made from the original negative. A comparison spectrum of titanium and vanadium lines (not shown here) provided the means for accurate determination of wavelengths and hence of identifications with neutral and ionized elements in the source. As shown along the upper margin of the spectrum, emissions of nitrogen

Figure 6.3 Slitless spectrum of lightning obtained in 1960
with transmission grating in front of large aerial camera

(neutral N I and ionized N II) predominate, with lesser contributions from oxygen and hydrogen. Hydrogen constitutes less than 0.0001 percent of dry air by volume, but since lightning occurs in an atmosphere saturated with water vapor, atomic hydrogen is plentiful as a result of the nearly complete dissociation of water by the electrical discharge. *(Photo courtesy of Lowell Observatory.)*

The early history of lightning spectroscopy is reviewed, with detailed references, by Salanave (1961) and by Orville and Salanave (1970). Also see reviews by Uman (1969a, Chap. 5) and Orville (1977b).

Slitless spectra of lightning, photographed with a large aerial camera and a transmission diffraction grating (Figs. 6.3-6.5). The spectrum in Figure 6.3, photographed during the summer of 1960 in Tucson, Arizona, initiated what may be called a new era of lightning spectroscopy. Great progress had been made in the previous decade (especially by the Bausch and Lomb Optical Company in the United States) in the production of transmission diffraction gratings by replication in thin plastic films mounted on glass. These gratings have grooves shaped so that over half of the incident light is concentrated (*blazed*) into the first order spectrum on one side of the undeviated beam, or zero order. The old, unblazed gratings had put most of the incident light into the zero order, where it was wasted. The new gratings could be obtained in sizes large enough to cover the aperture of a large camera lens — the F/6 Aero Tessar, for instance, which had already proven so effective in direct photography of lightning (see Fig. 4.1).

The great improvement shown in Figure 6.3 over slitless spectra previously obtained is due to the excellent telephoto characteristics of the camera lens and the comparatively large relative aperture of the optical system: f/8 with a focal length of 61 centimeters. Kodak Royal Pan film recorded the spectrum from approximately 3800 to 6200 Å. The bright spectral image of the lightning channel at the far right is at 6157 Å and is due to neutral atomic oxygen (O I). The sharp image at the far left is at 3995 Å, from singly ionized nitrogen (N II). At still shorter wavelengths there appear faint, diffuse emission bands near 3840, 3883, and 3914 Å, respectively. These emissions are of molecular origin and are quite variable from one flash to the next. The first two are due to CN and the third is from N_2^+. Clearly, the integrative property of the photographic film has recorded luminous outputs from both high and low temperature phases of the electrical discharge in air. (See Fig. 6.6 for more detailed identifications of the principal emission features in this region of the spectrum of lightning.)

Figure 6.3 shows the entire length of the flash below the cloud base; it is approximately 3 kilometers long. Horizontal streaks show the continuous spectrum, enhanced at points where the discharge may be oriented in the line of sight or along the direction of dispersion. In some other cases, it may be that the channel is intrinsically brighter at some points, or longer-lasting — as in bead lightning (see Fig. 3.7).

Amateur photographers can take colorful pictures of the spectrum of lightning similar to the illustration at the beginning of this chapter. All one need do is obtain a plastic replica transmission grating, and tape it in front of the camera lens. These diffraction gratings are inexpensive and can be purchased from suppliers of physics laboratory equipment. They are usually sold already mounted in a 2″ x 2″ slide.

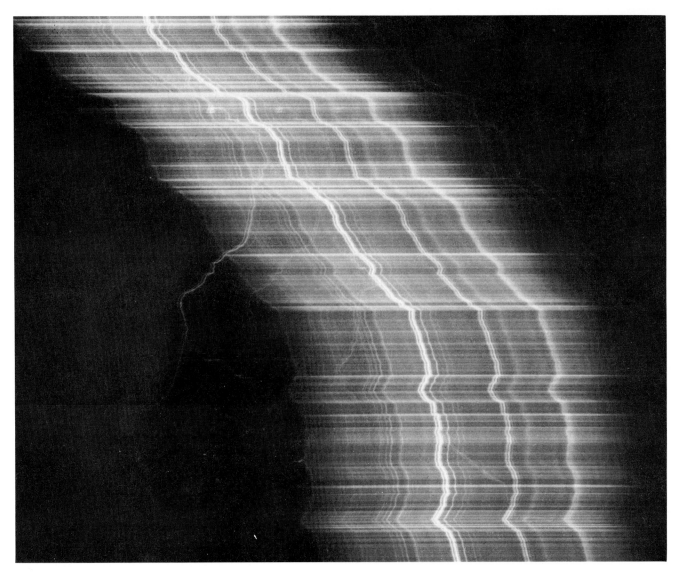

Figure 6.4 Slitless spectrum showing more resolution of detail than in Figure 6.3

The grating must be mounted on the front ring with dispersion in the horizontal direction. To determine this direction, simply look through the grating at a distant bright light and notice which way the colors are spread out on both sides. Then, after mounting the grating, follow the procedure already described for Figures 2.20 – 2.22 but set the lens at F/4.5 or F/5.6. Ektachrome 200, or a film with similar properties, is recommended. Also, the view finder should be centered about half a frame to the left or right of where the photographer is seeing lightning flashes.

Figure 6.4 shows a slitless spectrum obtained with a more critically focused lens than for the preceding illustration; the sharper definition of the emission lines is

readily apparent, and the obvious importance of having the best obtainable lens for this kind of photography is well illustrated. The section of lightning channel shown here is about 2000 meters long. The spectrum was photographed on panchromatic film through a yellow filter (Wratten #12). Wavelengths shorter than 5200 Å are cut off on the left; the other end of the spectrum extends to almost 6200 Å, the same as for Figure 6.3.

The photograph shown in Figure 6.5 was obtained in 1961, during the second summer of lightning spectrum photography at the University of Arizona in Tucson. The same camera, objective grating, and film were used as for Figure 6.3, but this flash produced a stronger exposure on the negative. The molecular bands

Figure 6.5 Molecular emission bands due to CN and N_2^+ around lightning flash

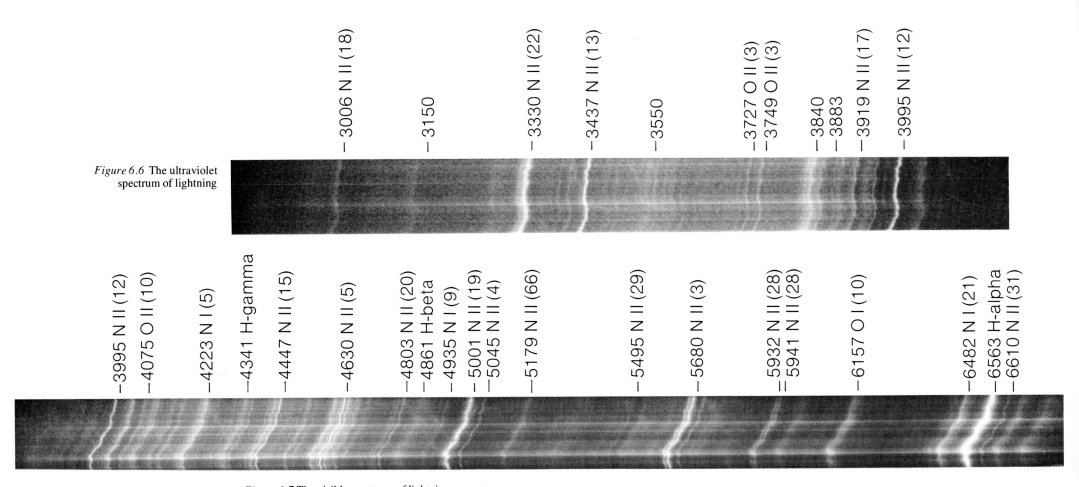

—3006 N II (18)
—3150
—3330 N II (22)
—3437 N II (13)
—3550
—3727 O II (3)
—3749 O II (3)
—3840
—3883
—3919 N II (17)
—3995 N II (12)

Figure 6.6 The ultraviolet spectrum of lightning

—3995 N II (12)
—4075 O II (10)
—4223 N I (5)
—4341 H-gamma
—4447 N II (15)
—4630 N II (5)
—4803 N II (20)
—4861 H-beta
—4935 N I (9)
—5001 N II (19)
—5045 N II (4)
—5179 N II (66)
—5495 N II (29)
—5680 N II (3)
—5932 N II (28)
—5941 N II (28)
—6157 O I (10)
—6482 N I (21)
—6563 H-alpha
—6610 N II (31)

Figure 6.7 The visible spectrum of lightning

referred to above are more apparent in this spectrum. *(Photos courtesy of Institute of Atmospheric Physics, University of Arizona, Tucson.)*

The photographic spectrum of lightning, from ultraviolet through infrared (Figs. 6.6-6.8). These three slitless spectra, with principal emission lines identified by wavelength (in angstroms) and chemical origin, cover almost the entire range accessible to photography. The short wavelength limit of the ultraviolet on the left of Figure 6.6 is set by atmospheric absorption — principally by ozone generated in the air around the discharge and in the ground layer during a thunderstorm. A special ultraviolet passband filter that eliminates almost all of the visible spectrum cut off the spectrum on the right of the photograph. A quartz-fluorite achromatic lens was used with a quartz-base diffraction grating so that the absorption properties of typical glass optical

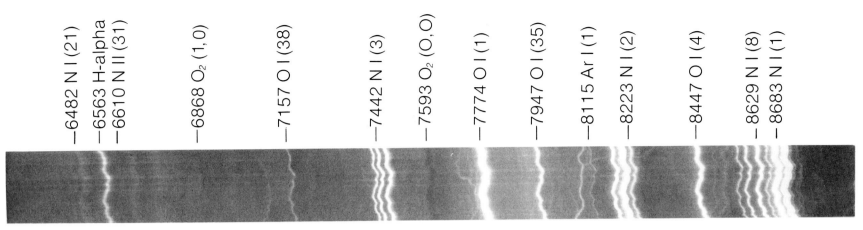

—6482 N I (21)
—6563 H-alpha
—6610 N II (31)
—6868 O₂ (1,0)
—7157 O I (38)
—7442 N I (3)
—7593 O₂ (O,O)
—7774 O I (1)
—7947 O I (35)
—8115 Ar I (1)
—8223 N I (2)
—8447 O I (4)
—8629 N I (8)
—8683 N I (1)

Figure 6.8 The red and infrared spectrum of lightning

components were avoided. Figure 6.7 displays what may be regarded as the entire visible spectrum of lightning, from violet to red. It was recorded on Kodak Tri-X Arecon, a panchromatic film with extended sensitivity to red light. The long wavelength limit in infrared, shown in Figure 6.8, was determined by the spectral sensitivity of the photographic film — in this case, Kodak High Speed Infrared.

The spectra in Figures 6.6-6.8 are of different flashes and are comparable only insofar as they are typical of time-integrated spectra — that is, those in which the radiation from all component strokes makes up the total exposure. Numerals in parentheses above lines of atomic origin identify the corresponding multiplet number for that element, as listed in "A Multiplet Table of Astrophysical Interest" by C. E. Moore (1945). In the case of lines and bands of molecular origin, as at 6868 and 7593 Å, the numbers in parentheses identify the vibrational energy level transitions that give rise to these spectral features. (*Photos courtesy of Institute of Atmospheric Physics, University of Arizona, Tucson.*)

Detailed descriptions of the spectrum of lightning, with references to other work, can be found in Uman (1969*a*, Chap. 5, which contains an extensive list of emission lines and bands extending from 3159 to 9393 Å); also in Orville and Salanave (1970), and Orville (1977*b*).

Sections of the red and infrared spectrum of lightning, enlarged from well-exposed and exceptionally sharp negatives of slitless spectrograms (Figs. 6.9-6.11). These photographs show complex structures along the channel as well as fine details of the emission lines and absorption bands. The three spectra were all obtained during the same lightning storm in July 1965 with the Aero Tessar telephoto lens and grating used to photograph the spectrum in Figure 6.8. Kodak High Speed Infrared film was used to extend the spectral range out to nearly 9000 Å. The photographs were taken within a few minutes of each other, and all flashes were approximately 15 kilometers from the camera. The vertical length of the channel shown in each photograph is about 300 meters.

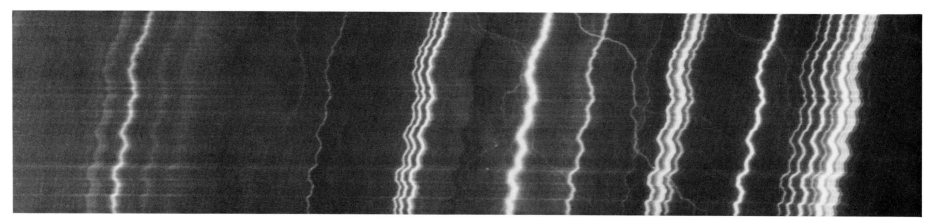

Figure 6.9 Enlarged detail from Figure 6.8

The reproduction in Figure 6.9 shows a larger vertical section, and more detail, of the same negative used to produce Figure 6.8. Tortuosity of the channel is resolved into elements that are about 10 meters in length.

The significant feature of Figure 6.10 is the very strong exposure of the continuous spectrum, which provided a background for *absorption bands* due to molecular oxygen and water vapor. The complexity of these absorptions, which have the same atmospheric origin as the telluric bands of the solar spectrum, is well shown.

In sharp contrast to Figure 6.10, the spectrum in Figure 6.11 was lightly exposed and the channel very tortuous. No simultaneous direct photograph is available to verify the nature of the whole flash, but it is probable that this was a branch rather than a main channel, and therefore the exposure was weaker. This photograph shows the scale of tortuosity on a significantly smaller scale than in Figures 6.9 and 6.10 — down to dimensions of only 3 or 4 meters, photographed from a distance of approximately 15 kilometers. The infrared region of lightning's spectrum, with its relatively weak continuum, provides well-resolved lines that are members of the same multiplet and exhibit a range of intensities that can be calculated from the atomic transitions involved. Thus there is a series of steps in the exposure of the discharge channel, and a particular detail (bead, knot, line-pair, etc.) can be examined on the image that shows it to best advantage. One millimeter on this reproduction equals approximately 4 meters at the lightning. *(Photos courtesy of Institute of Atmospheric Physics, University of Arizona, Tucson.)*

Qualitative and quantitative aspects of the infrared spectrum of lightning are discussed by Salanave (1969a).

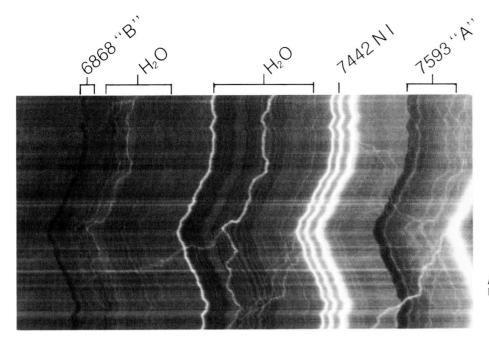

6868 "B" H₂O H₂O 7442 N I 7593 "A"

Figure 6.10 Continuous spectrum with absorption bands of atmospheric O₂ and H₂O

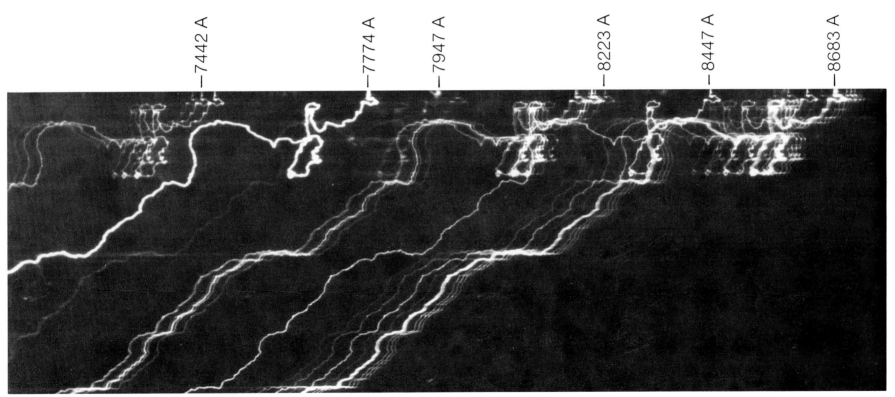

7442 A 7774 A 7947 A 8223 A 8447 A 8683 A

Figure 6.11 Slitless spectrum of lightning showing channel tortuosity on a small scale

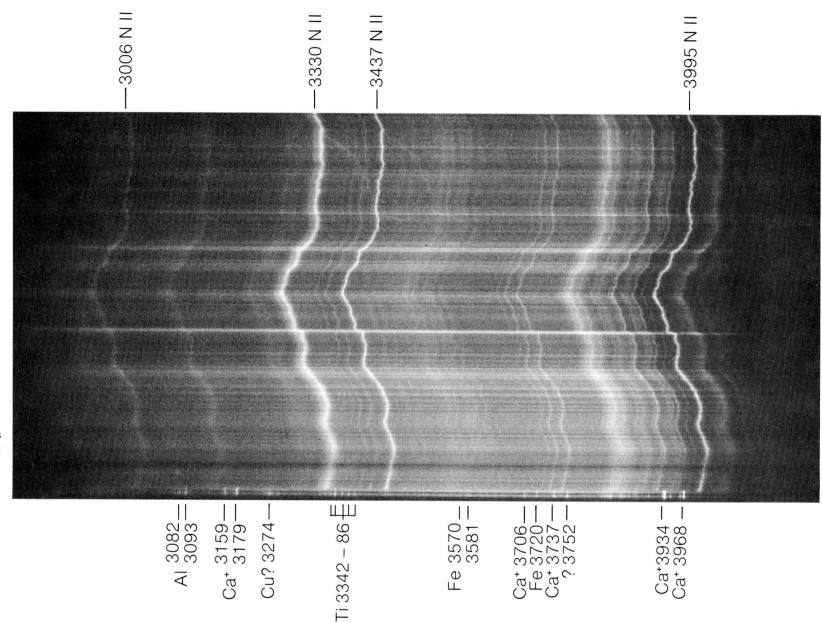

Figure 6.12 Earth lines in spectrum of lightning

Earth, or pole, lines in the spectrum of lightning, emitted at the point where the flash made contact with the ground (Fig. 6.12). Figure 6.6 was taken from a section in the upper half of this spectrum. An unusual feature at the bottom of the spectrum, shown here, is the appearance of emission lines due to metals vaporized at the point where the discharge struck. The prominent emissions of aluminum, iron, and calcium are to be expected. The appearance of titanium may be surprising until one recalls that titanium is relatively common in the crust of the earth. It is approximately 15 percent as abundant as either calcium or iron, and nearly 150 times as plentiful as copper. One of copper's most sensitive indicators is the emission at 3274 Å; it appears faintly in this spectrum. One might suspect that copper is more likely to be found in the ground of southern Arizona than almost anywhere else!

This spectrum was obtained with a quartz transmission diffraction grating and a quartz-fluorite doublet lens with a focal length of 250 millimeters and an aperture of F/9. No good estimate of the distance to the flash was obtained, but on the reasonable assumption that it was between 5 and 10 kilometers away, the apparent height of the earth lines indicates that the metallic ions migrated upward along the channel a distance of 10 to 20 meters during the brief time the discharge current was flowing. *(Photo courtesy of Institute of Atmospheric Physics, University of Arizona, Tucson.)*

Lightning channels photographed in light emitted by the H-alpha line of hydrogen (Figs. 6.13, 6.14). The first member of the Balmer series in the spectrum of hydrogen has a wavelength of 6563 Å, that is, in the

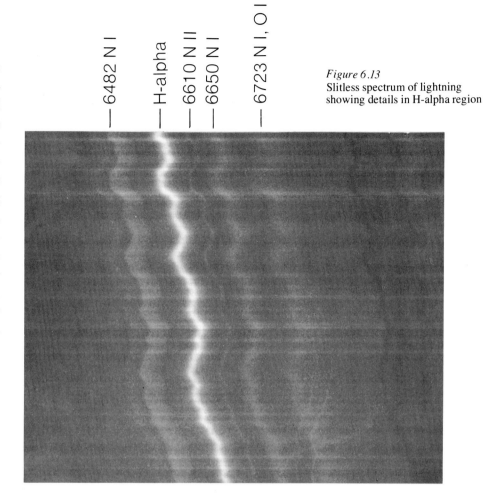

6482 N I | H-alpha | 6610 N II | 6650 N I | 6723 N I, O I

Figure 6.13
Slitless spectrum of lightning showing details in H-alpha region

deep red. Ordinary panchromatic type films are not sensitive to this part of the spectrum, but emulsions with sensitivity to longer wavelengths, including the near infrared, show that the H-alpha emission is a prominent feature of lightning's spectrum. The atomic hydrogen is derived from water vapor dissociated by the electrical discharge.

The enlarged section of the slitless spectrum shown in Figure 6.13, exposed on Kodak High Speed Infrared

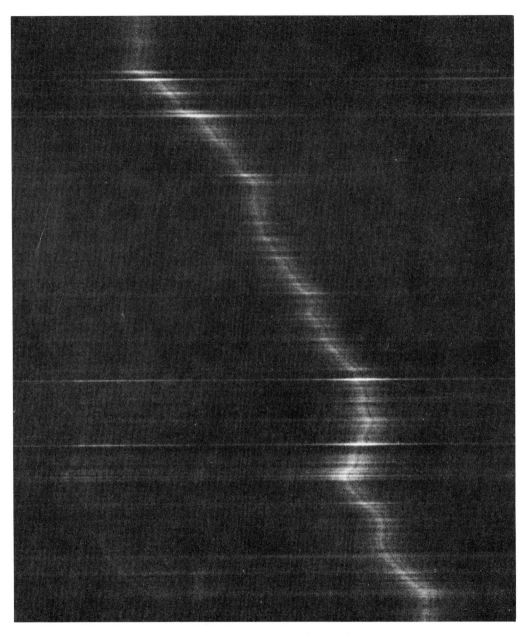

Figure 6.14 Detail of lightning channel photographed in H-alpha radiation

Increasing wavelength ⟶

with a 61-cm focal length lens, was printed to bring out the subtle details along and adjacent to H-alpha. The vertical section of channel shown here is about 300 meters long, and the smallest distinguishable sections along the channel are approximately 10 to 20 meters in length — the mesoscale of tortuosity.

The width of the H-alpha line, measured in angstroms, is sensitive to the electron density of the emitting gas. This broadening is due to the *Stark effect*; that is, the spread of the line-profile is a measure of the number of electrons per cubic centimeter in the plasma of the electric discharge. Quantitative analyses of these widths are best performed with spectra recorded on moving film, in other words, time-streaked spectra (see Figs. 6.17-6.19). Three diffuse emissions on either side of H-alpha are identified with neutral nitrogen (N I) and oxygen (O I); one sharp line is due to ionized nitrogen (N II).

Figure 6.14, a detailed picture of approximately 150 meters of lightning channel, was photographed in H-alpha radiation with a larger camera and grating than were used for the spectrum in Figure 6.13. The film was Kodak Shellburst, a special emulsion with extended red sensitivity and high acutance. *Acutance* is the term used by photographers to express the property of a film that enables it to record the subtle details in an image. In general, high-acutance films have typically high contrast and low graininess, although acutance is more than just a simple combination of these two properties.

The focal length of the lens was 122 centimeters; the flash occurred about 13 kilometers from the camera.

The dispersion (horizontal scale) on this reproduction is 12 angstroms per centimeter. The projected height (vertical scale) is about 10 meters per centimeter, with a 20 percent uncertainty due to the estimate of 13 kilometers for the distance to the flash. In any case, the resolution of structure along the channel is on the order of 2 or 3 meters — the microscale of tortuosity.

What appear as fine, dark features superimposed on the H-alpha emission are faint absorption lines that are members of an extensive, complex system of spectral bands due to water vapor in the light path between the lightning and the spectrograph. The broad, bright, horizontal spikes along the channel may be simply effects of greater exposure where the channel has a sharp bend with a short section of it lying in the line of sight or along the direction of dispersion (compare Figs. 6.3, 6.4). On the other hand, there may be sections of the channel (beads) where the luminosity, or the electron density, is markedly greater than elsewhere. The first circumstance could account for an increase in effective exposure and hence an apparent broadening. The second condition could produce an enhanced Stark effect and a real line-broadening. If both conditions were to occur together, the broadening of the line would be even more noticeable at these places. *(Photos courtesy of Institute of Atmospheric Physics, University of Arizona, Tucson.)*

Figure 6.14 and other photographs representing results obtained with high-resolution spectrographs applied to lightning have been discussed by Salanave (1969*a*).

Direct photograph of lightning correlated with the spectrum of the same section of discharge channel (Fig. 6.15). This pair of pictures compares the direct image of a lightning channel with its slitless spectrum. Figure 6.15*a* shows a flash at a distance estimated to be 6.5 kilometers. The photograph was obtained with a tele-photo lens of 122 centimeters focal length and is reproduced here on a vertical scale of approximately 25 meters per centimeter.

The spectrum of the same flash is shown in Figure 6.15*b*, reproduced with the same vertical scale as the direct photograph. The wavelength range is from approximately 5000 to 6000 Å; to identify the spectrum lines, see the central section of Figure 6.7. Notice the enhancement of the continuous spectrum where the channel bends in the direction of dispersion, as at the place about 25 millimeters from the lower edge of the spectrum. Also note, at a point 65 millimeters up, where the channel appears to be thicker and more luminous. This may be a place where the channel is oriented along the line of sight. *(Photos courtesy of Institute of Atmospheric Physics, University of Arizona, Tucson.)*

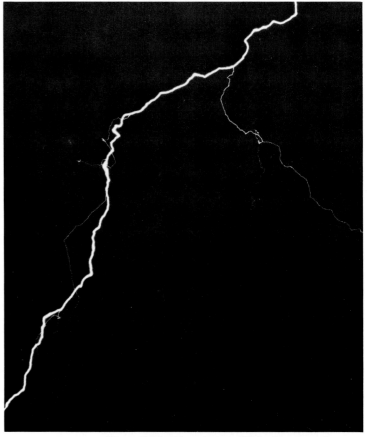

Figure 6.15a Direct image of lightning channel

Spectra of three consecutive strokes in a lightning flash, photographed on a rotating film drum placed behind a horizontal slot that isolated a short section of the discharge channel (Fig. 6.16). Figures 4.1-4.5 show that a lightning flash is typically made up of several component discharges called strokes. By moving either the camera or the film, it is easy to separate the images of individual strokes — at least when they are spaced in time by 10 milliseconds or more. But, until 1961, all spectra of lightning had been obtained with a fixed film

in order to build up an exposure of sufficient strength to produce a photographic image. Dispersing the lightning flash into a spectrum so reduces the effective exposure that it was generally not considered possible to get an adequate photograph if a time-streak, or image motion, further weakened the light falling on the film. However, H. Israel and G. Fries provided a noteworthy exception to this view when, in 1954-55 at the West German

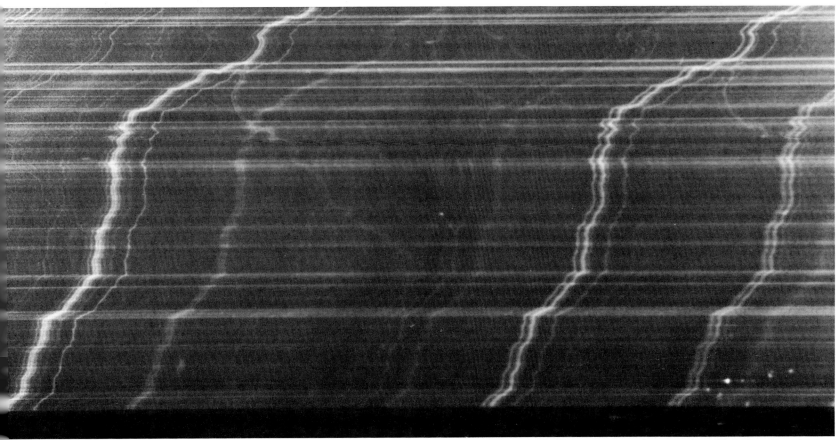

Figure 6.15b Slitless spectrum of same section of lightning channel

Meteorological Observatory in Aachen, they constructed and tested a time-resolving spectrograph for lightning photography. Their instrument consisted of a slitless spectrum camera in which a rapidly spinning disc with radial slots scanned the lengths of the spectral images of the lightning channel. The width of the slots and the rate of rotation were such that the time-resolution was 20 *microseconds* near the rim of the spinning disc.

This was a bold experiment and deserved to succeed, but the sought-for time-resolution was too high for incident light levels and available film speeds. Also, even if a well-exposed spectrum had been obtained, it would have been difficult to interpret because each time-element would have corresponded to a different section of the discharge channel; that is, the variables of time and position would have been mixed together.

Then, in 1960 at the University of Arizona, experience gained in photographing the spectrum of lightning with large cameras, fast films, and the new blazed diffraction gratings made it clear that separate lightning strokes should be bright enough to record their spectra on moving film (see Figs. 6.3-6.5). The rate of film motion need only be sufficient to separate the spectra of individual strokes; 1 centimeter in 20 milliseconds would do it.

The first device to photograph spectra of separate strokes consisted of a cassette in which a rotating drum carried a strip of film behind a horizontal slot that was placed just in front of the focal plane of the spectrograph. The drum was 18 centimeters in diameter and rotated at 60 rpm (once per second). The slot was 10 millimeters wide, 22 centimeters long, and served to isolate a vertical section of the lightning to be photographed. Spectra produced by strokes that occurred at least 18 milliseconds apart were recorded separately on the film. For example, Figure 6.16 shows the spectra of three consecutive strokes in a flash of lightning. The first one is at the bottom of the series, and the time intervals between them are 40 and 52 milliseconds, respectively. The length of the isolated section of lightning channel is about 100 meters. *(Photo courtesy of Institute of Atmospheric Physics, University of Arizona, Tucson.)*

The importance of this new technique lay in the fact that for the first time the spectra of individual strokes in a flash could be recorded and analyzed separately. Previous spectra of lightning flashes photographed on fixed film were typically the superposition of several strokes. The cumulative result of an intermittent exposure, however, is not simply the sum of the individual exposures: the discrepancy depends in a complicated way upon the duration and interval between the exposures. Unlike a strobe light, the intermittency of a multistroke lightning flash is quite irregular, which adds to the difficulties of quantitative analysis.

Time-resolving a lightning flash produces individual spectra that can be analyzed by comparing them to a standard light source having nearly the same duration and relative spectral intensity as a typical lightning stroke. A xenon flash lamp gives a good approximation for this purpose. After a calibration standard has been set up, relative emission line intensities can be determined from the spectrum; these in turn are a measure of the temperature of the air made incandescent by the electric discharge. The lines emitted by ionized nitrogen (N II) are best suited for this measurement in lightning; the first results from spectra obtained at the University of Arizona were in the range of 24,200 to 28,400 degrees Kelvin. Later studies showed that the average value of these estimates was probably within 10 percent of the maximum temperature reached in a typical stroke. That is to say, lightning's temperature peaks at nearly 30,000 degrees Kelvin for a microsecond or two at the beginning of the great surge of current that heats and ionizes the air in its path.

The first analysis of flash-resolved spectra to find the temperature of individual lightning strokes was made by Prueitt (1963). The construction of the rotating drum cassette, and some of the first results obtained with it, are described by Salanave et al. (1962). A more detailed discussion of the theory of temperature determination from emission spectra, and a critical interpretation of the results, are presented in Uman (1969a, Chap. 5).

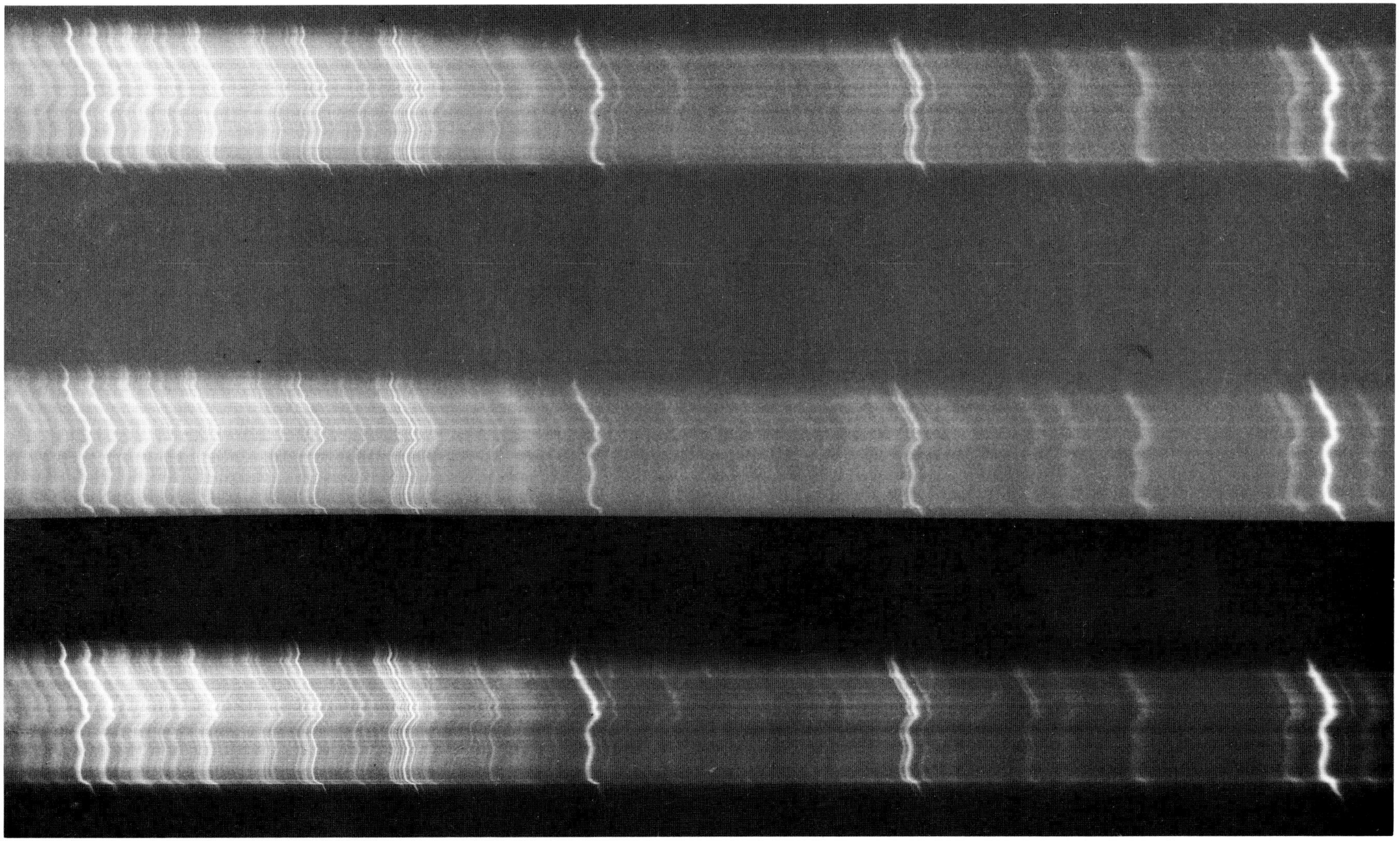

Figure 6.16 Spectra of three consecutive strokes photographed on rotating film drum

3995 N II

5679 N II

6563 H-alpha

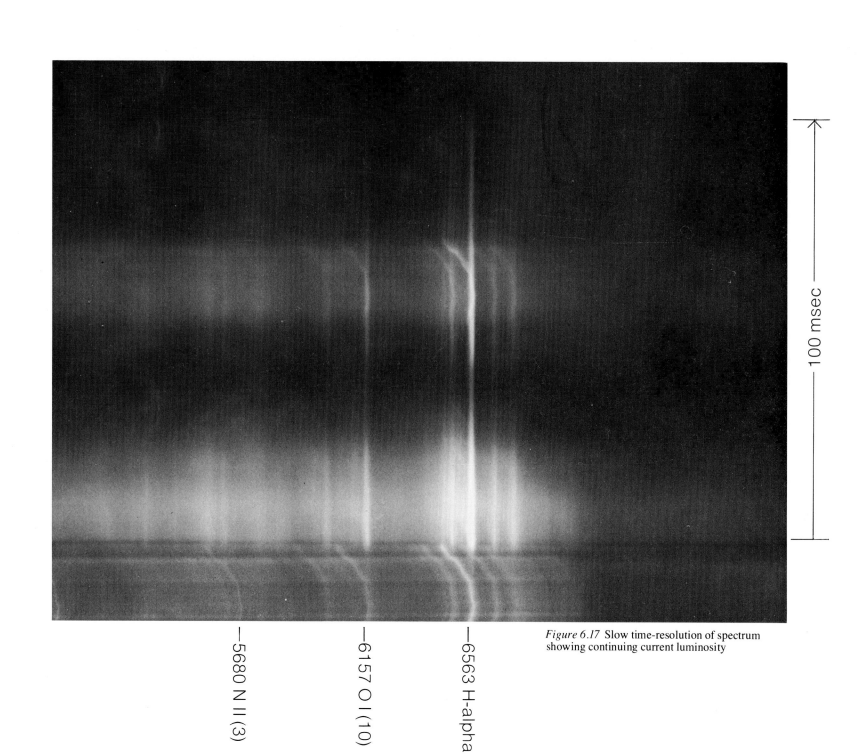

100 msec

5680 N II (3)

6157 O I (10)

6563 H-alpha

Figure 6.17 Slow time-resolution of spectrum showing continuing current luminosity

Persistence of spectrum lines following a lightning stroke, shown with slow time-resolution (Fig. 6.17). This spectrum shows the effect of continuing current luminosity, the same phenomenon that extends the image of one or more strokes in a flash when it is photographed with a moving camera (see Fig. 4.5). The result is a streaking out of the spectrum beyond the few tens of microseconds during which the stroke is near its peak of brightness. For this illustration, the spectrum of the stroke (at the bottom of the figure) was printed heavily (dodged), to compensate for its excessive density on the negative. Thus the spectrum of the stroke can be compared in the print with that of the continuing current luminosity with respect to the identity of the emission features, but not their relative intensities.

The most prominent feature is the persistence of H-alpha at 6563 Å; it can be followed for 0.1 second after the stroke. Notice that the luminosity almost ceases at 40 milliseconds into the continuing phase, and then starts up again without reaching its previous level of brightness. The same sequence is true for other features in the spectrum, but with differences in degree. For example, while the N II (ionized nitrogen) line at 5680 Å does not persist into the continuing luminosity phase, the neutral oxygen (O I) at 6157 Å does, almost as prominently as H-alpha. This corresponds closely to the fact that N II requires a higher temperature to be produced in air and to emit its spectrum than either neutral oxygen or hydrogen. In this qualitative view, temperature is the most important factor to consider for an understanding of what has taken place. The four diffuse continuing emissions on either side of H-alpha can, even without specific identification, be confidently attributed to neutral atoms. They are in fact due to N I and O I (see Fig. 6.13). *(Photo courtesy of Institute of Atmospheric Physics, University of Arizona, Tucson.)*

The interpretation of this rather unique spectrum has been discussed in more detail by Orville and Salanave (1970).

Persistence of spectrum lines in a lightning stroke, shown with fast time-resolution (Figs. 6.18, 6.19). The photographic separation of the spectra of individual strokes in a flash was an important first step in applying spectrum analysis to determine physical characteristics of the lightning channel, such as temperature, electron density, and pressure (see Fig. 6.16). However, in this procedure the individual spectra are exposed for the entire high-luminosity phase of the stroke, which lasts 100 to 200 microseconds. Thus the resulting spectra represent a time-averaged temperature, electron density, and other physical parameters taken over the luminous lifetimes of the individual strokes.

For the measurement of rise times, peak values, and decay times within individual strokes, it is necessary to have time-streaked spectra (and any other data to be correlated with them) speeded up to a scale of microseconds. With this increase in time-resolution, a smaller section of channel must be isolated, so the measurements refer to a short length of the lightning channel (on the order of 10 meters), and the time required for the stroke to traverse the length is small (a microsecond or less). At first it might seem to be simply a matter of speeding up the drum used to resolve the flash spectrum and narrowing the channel isolator slot to achieve the desired result. However, on second thought it is clear that the film drum would have to rotate at least a hundred times faster, and the slot be made about one tenth

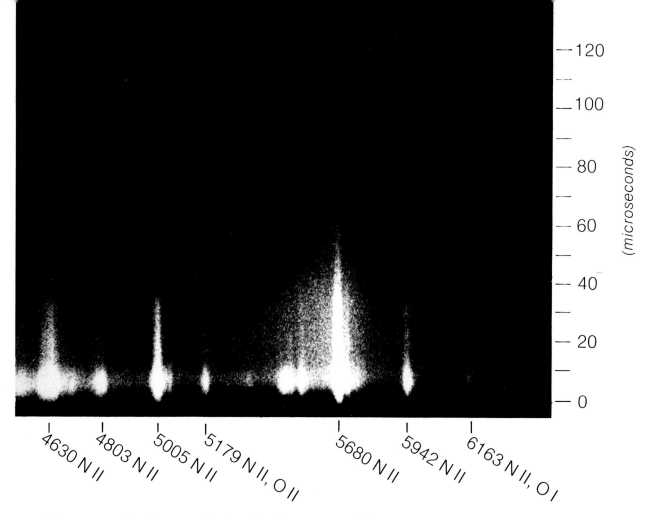

Figure 6.18
Fast time-resolution of spectrum

as wide, to get the time-resolution in the range of the few microseconds required for analysis of a single stroke. The drum on which a flash was resolved into its component strokes had film wrapped on the outside, which presented no problem at 60 rpm. But at 6000 rpm it would be impossible to keep the film on the drum because centrifugal force would throw it off. A different approach was required.

The first photograph of the slitless spectrum of a lightning stroke with time-resolution measured in microseconds, shown in Figure 6.18, was obtained in 1965 by R. E. Orville at the University of Arizona. The preliminary results of this outstanding advance in the optical study of lightning were published by Orville (1966). A film strip was located on the *inside* of a stationary drum cassette, and the spectrum was swept around by a three-sided mirror rotating at 3,000 rpm, that is, 50 times a second. The optical system, combined with a narrow horizontal slot, isolated approximately 10 meters of the lightning channel with a time-resolution of 5 microseconds. A very fast film had to be used; in this case it was Agfa Isopan Record (ASA 1200). Because the emulsion was not sensitive to the red end of the spectrum, the H-alpha line is not shown in the photograph.

This spectrum shows that the first emissions to appear are those of ionized nitrogen (N II), followed in a few microseconds by the continuum and emission lines from neutral atoms. Temperature determinations from this and other spectra with fast time-resolution show the peak value is reached very quickly, in less than 10 microseconds. In some instances, the measurements did not include the peak; the temperature simply decreased from some initial high value. Typically the peak temperatures (or those highest observed) were in the range of 28,000 to 31,000 degrees Kelvin.

The time-streaked spectrum of a lightning stroke shown in Figure 6.19 was obtained in very much the same way as the previous one. A Dynafax camera, with a stationary mirror and spinning drum, photographed the spectrum of a 10-meter section of channel with a time-resolution of 2 to 5 microseconds. Using Kodak Recording 2475, an extremely fast film sensitive to the deep red end of the spectrum, the region around H-alpha was also recorded. Notice that the hydrogen line apparently reaches its peak at the same time as the continuum; that is, after the lines due to ionized nitrogen (N II) have already appeared. This corresponds to the fact that H-alpha does not require as high a temperature as N II for excitation and emission. Therefore, H-alpha appears later and persists longer than, for example, the lines at 5680 and 5942 Å.

An important result based on this spectrum, and others showing the H-alpha emission with microsecond time-resolution, was the determination of electron density (ionization) as a function of time. This was done by measuring the width of the H-alpha line at different

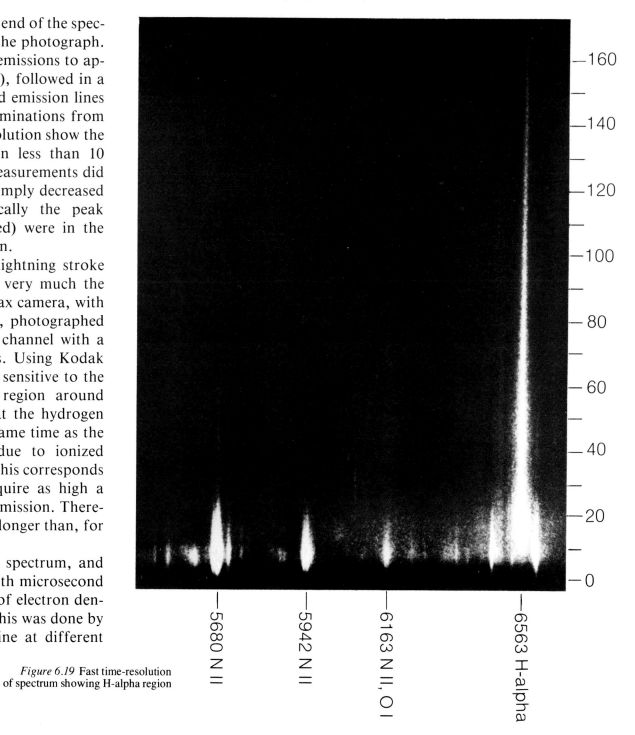

Figure 6.19 Fast time-resolution of spectrum showing H-alpha region

Figure 6.20 Spectrum of long spark over water

positions along its time-streak and calculating the electron density from the interatomic Stark effect, to which the hydrogen spectrum lines are particularly sensitive. The results of such analyses indicate that electron density rises to 10^{18} per cubic centimeter in 2 microseconds or less, then drops to 10 percent of that value in about 30 microseconds, after which it gradually goes lower as the lightning channel cools down in the next 100 microseconds or so. *(Photos courtesy of R. E. Orville, State University of New York at Albany.)*

After the first report about these new data on lightning in 1966, additional results were published by Orville (1968a); see also Uman (1969a, Chap. 5). A later discussion, with observations, analyses, and an updated bibliography, can be found in Orville (1977b).

The spectrum of a long spark in air above a water surface (Fig. 6.20). The spark, whose undispersed image appears on the left, was 5 meters long and was produced by a 5-million-volt impulse generator at the Lightning and Transients Research Institute, Miami Beach, Florida. The upper electrode consisted of a copper rod suspended above a tank of water, its polarity negative with respect to the water. This polarity simulated the negative charge which thundercloud bases normally carry under lightning conditions.

The spectrum, produced by a transmission diffraction grating placed in front of the camera lens, extends from 3995 Å, the first strong image on the left, to 6563 Å, the H-alpha line on the far right. The sharp cutoff of the spectrum beyond H-alpha is due to the rapid drop in sensitivity of the Kodak type M photographic plate used for the exposure. The spectrum drops off less abruptly at the violet end because of the gradual increase in absorption of ultraviolet light by the glass optical system.

The range of wavelengths here is the same as in Figure 6.7, to which the reader should refer for identification of other lines. To aid this identification, it should first be noted that prominent spark emissions due to N II at 4630, 5001, and 5680 Å are 18, 28, and 47 millimeters to the right of 3995 Å, respectively. A narrow spectrum of emission lines appears faintly across the top. These are pole lines at the tip of the negative electrode, from which a trace of copper, with some impurities, has been evaporated (compare with Fig. 6.12).

The path of the spark splits into three distinct chan-nels within a section extending from 0.6 to 1.3 meters above the water surface. This braiding, or looping, effect sometimes occurs in that section of the channel where a downward leader and the upward streamers meet to initiate the long spark. This multiplicity of the discharge channel is commonly observed in long sparks when they are photographed with high spatial resolution. In this case, the spark was 16 meters from the camera, which was equipped with a lens of 180 millimeters focal length. The vertical scale of the reproduction is approximately 10 millimeters to 40 centimeters at the spark. The braiding effect is rarely observed in lightning because the ground strike point is usually too remote for the camera to capture such fine detail (see, however, Figs. 7.1 and 7.2). *(Photo courtesy of Institute for High-Tension Research, University of Uppsala, Sweden.)*

A comparison of the spectrum of lightning with that of a long spark (Fig. 6.21). These illustrations show the similarities and differences in the spectra of lightning and a long spark, as recorded by the same slitless spectrograph. The upper spectrum (a) is from a 900-meter section of a lightning channel whose visible length was about 3000 meters. The lower spectrum (b) was obtained from a spark 4.5 meters long, produced by the Westinghouse high-voltage generator at Trafford, Pennsylvania. As in Figure 6.20, the upper electrode was a thin rod with negative polarity. The lower electrode was a grounded metal plate. This reproduction includes 80 percent of the spark gap; neither electrode is shown in the picture.

In this figure the spectra have been reproduced with the upward and downward directions reversed. No sys-

| 3995 N II | 4447 N II | 4630 N II | 5001 N II | 5179 N II | 5680 N II | 5935 N II | 6158 O I |

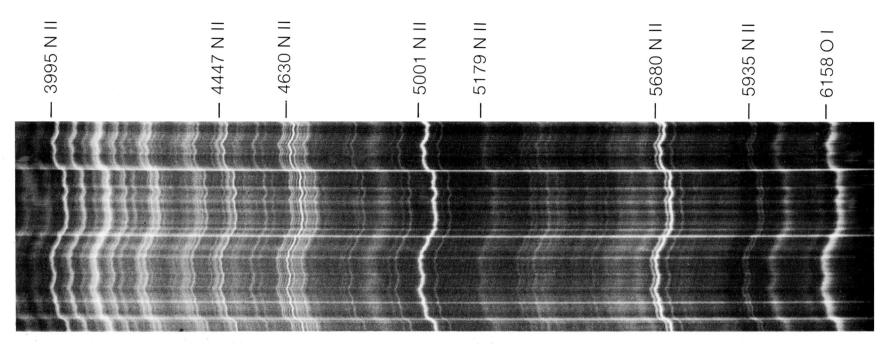

Figure 6.21a Spectrum of lightning

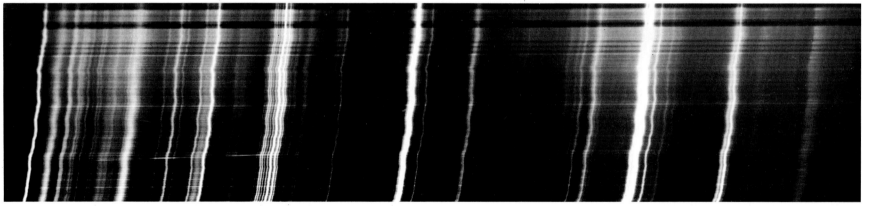

Figure 6.21b Spectrum of long spark

tematic cloud-to-ground variations are apparent in the photograph of lightning; less than one-third of the flash is represented. But in the case of the spark, some emission lines clearly broaden toward the positive pole of the discharge, while others show this tendency only slightly.

Both spectra were photographed with the same Aero Tessar F/6 camera lens, which has a focal length of 61 centimeters. The transmission diffraction grating in front of the lens has 600 grooves per millimeter and gives a spectral dispersion on the film of approximately 25 angstroms per millimeter. The effective resolution is 1 or 2 angstroms, depending on the focus adjustment, film graininess, and other instrumental factors.

The distance to the lightning flash was 25 kilometers, but the spark was only 100 meters from the camera. The *spatial resolutions* (vertical scales) differ by a factor of at least 250, and, therefore, nothing can be deduced from these photographs about the comparative structures of the channels as to thickness or tortuosity. On the reproduction of the spark spectrum the vertical scale is 1 millimeter to 7 centimeters, and details separated by only a few millimeters can be resolved on the original negative. The *spectral resolutions,* however, are nearly the same. On these reproductions the wavelength scale is about 10 angstroms per millimeter. Individual lines can be found and identified with the same confidence in either spectrum, except in cases where broadening of the lines results in some loss of resolution. The diameter of the spark has been estimated from selected spectral images; the results range from 1 to 2 centimeters, depending upon which line is used, whether the emission originates from neutral oxygen or ionized nitrogen, and how sensitive it is to the effect of Stark broadening.

The lightning and the spark were photographed on different films — Kodak Royal Pan and Linagraph Shellburst, respectively. Although the overall sensitivity of Shellburst is considerably above that of Royal Pan in the wavelength range covered by these spectra, the ratio of film sensitivities between 4000 and 6200 Å remains nearly the same, so that intrinsic similarities and differences in line intensities can be noted. For example, ionized nitrogen (N II) lines at 3995, 4630, 5001, and 5680 Å maintain about the same relative intensities in lightning and in the spark. However, the N II emissions at 5179 and 5935 Å appear to be weaker in lightning. On the other hand, the line of neutral oxygen (O I) at 6158 Å is clearly more prominent in lightning. Just to the left of the 5001-Å line in the lightning spectrum are two faint, narrow lines, and then a broad, diffuse emission. These features are absent in the spectrum of the spark, even though it is more strongly exposed. The reader may suspect — and rightly so — that these lines are emitted by neutral atoms. In fact, the sharp lines are due to N I, and the diffuse one is H-beta, the second member of the Balmer series of hydrogen lines. Comparisons such as these illustrate basic differences in temperature, ionization, electron density, duration, and other characteristics of lightning versus those of a long spark.

Lightning itself exhibits some interesting spectral variations from one flash to the next. A qualitative discussion of these phenomena has been published by Salanave (1964) and Meinel and Salanave (1964). When spectra of lightning and long sparks are carefully calibrated and analyzed quantitatively, a great deal of information about the physical characteristics of these sources emerges. The reader interested in quantitative details is referred to Uman (1971). *(Photographs: a, by the author, courtesy of Institute of Atmospheric Physics, University of Arizona, Tucson; b, by R. E. Orville, courtesy of Westinghouse Research Laboratory.)*

Violent lightning storm over desert terrain

Courtesy of University of Arizona Institute of Atmospheric Physics

LIGHTNING MISCELLANY

7

Lightning flash to the corner of a chimney, showing a multiple channel 9 meters above the point of ground contact (Fig. 7.1). This photograph was taken in 1960 on 35-mm film from a distance of only 66 meters, with a lens of 50-mm focal length and with the aperture stop at f/11. The photographer had his camera pointed, shutter open, toward a thunderstorm in the hope of catching distant lightning when this bright flash struck just outside his window after the camera shutter had been open only 30 seconds.

Braiding, or looping, of the channel is sometimes observed in photographs of long sparks (see Fig. 6.20), but it is seldom recorded in lightning because the flashes are rarely close enough to reveal such fine detail. In the case illustrated here, the braided section is about 4.6 meters long and at least 60 centimeters wide at its greatest separation. This unusual picture is an excellent illustration of what sometimes occurs when downward leaders and upward streamers meet just a few meters above ground to initiate return strokes. Golde (1967)

Figure 7.1 Lightning with multiple channel above the point of ground contact

used this photograph as an example of the leader-streamer junction process.

The halo next to the channel is of uncertain origin since the photograph was taken through a window pane, and there was probably rain and mist around the flash. In this respect, the photograph resembles Figure 2.14. See Figures 7.4-7.6, however, for a photograph of what may be a discharge corona associated with a lightning channel. *(Photo courtesy of Jouko Wesanterä, Helsinki, Finland.)*

Lightning flash to a tree, photographed at close range (Fig. 7.2). This photograph was obtained during a remarkable stroke of luck similar to that which produced Figure 7.1. During a series of lightning experiments carried out in 1967 near Lugano, Switzerland, a camera was used to photograph flashes striking distant mountains, particularly the top of mount San Salvatore (see Fig. 2.7). While the shutter was open, in readiness for whatever lightning might appear over the mountains, a flash hit a tree in the foreground, at a distance of only 60 meters. Although the camera was prefocused for infinity, the lens had such a short focal length (55 mm) and sufficiently small aperture (f/5.6) that the tree and the lightning above it were well within the range of critical focus. Thus the downward branch from the upper section of the channel and the bright upward streamer just above the tip of the tree are sharply defined, as are the clusters of leaves on the tree.

This reproduction from a Kodachrome transparency shows 30 meters of the lightning flash above the tree top, and another section apparently running down the trunk. There can be no doubt that the tree was actually struck, that is, this was not an accidental alignment with the camera. The branches on the right are in

Figure 7.2 Lightning flash
photographed at close range

the full glare of the flash (front lighted), while those on the left are in partial shade (back lighted), because the lightning ran along the right side of the tree trunk. A careful examination of the tree on the following day revealed no damage to the bark. In many cases, large trees are torn limb from limb by lightning, and it is remarkable that this strike left no visible mark on a tree only 7 meters high. *(Photo courtesy of R. E. Orville, State University of New York at Albany.)*

The junction point between faint downward branches and upward streamers occurred somewhere between 10 and 12 meters above the tree top (6 and 7 centimeters on this reproduction). Branches and streamers made electrical contact when the air broke down under the stress of several million volts. Then the return stroke propagated upward toward the base of the thundercloud in a great flash of light. A photograph showing upward streamers associated with lightning flashes to rocky ground is reproduced in Figure 7.3.

The large apparent diameter of the lightning channel (30 centimeters, as interpreted from the photograph) is an effect of overexposure on the film. The reddish halo in the colored original is also an exposure effect, since Kodachrome film makes a subject appear excessively red when it is overexposed. Unlike the example in Figure 7.4, there is no convincing reason, on the basis of this photograph alone, to suppose there is a real sheath of light surrounding the lightning channel.

This photograph, with additional measurements and more discussion, has been published by Orville (1968*b*).

Upward streamers from the ground near lightning strike points on mountainous terrain in southern Arizona (Fig. 7.3). This photograph shows two light-

ning flashes striking rocky ground that is partially covered with low shrubs, or chaparral. Two unconnected upward streamers can be seen at the base of the channel on the left; another streamer is barely visible next to the channel on the right. The nearer, brighter flash struck near the summit of Kitt Peak (on the left), on a slope at a distance of 810 meters from the camera. The other flash hit near the top of a neighboring summit, at a distance of 1565 meters. These distances could be accurately surveyed because the location of the camera was precisely known, and the small areas within which each flash struck could be identified from the photograph.

This exposure was made in 1972, on Kodachrome II film with a Nikkor 28-mm lens set at f/3.5. The shutter was open for about 2 minutes. Ordinarily, such a large aperture setting would not be recommended for lightning photography, as these greatly overexposed images of the two flashes illustrate. But in this case the wide-open lens caught the faint details of the associated streamers and branches and — of equal importance — recorded the ground features for later identification and distance determination.

This reproduction, enlarged 13 times, is scaled so that 10 millimeters on the print equals 22 meters at the nearer point of contact with ground. Due to the effect of light spreading in photographic images, diameters of lightning channels measured directly on uncalibrated film without the aid of a densitometer can be regarded only as upper limits. On this basis, the diameters of the upward streamers approach 1 meter, but they could be only half as wide. However, measured lengths of channels, branches, and other comparatively large features can be accepted with more confidence and a correspondingly smaller percentage estimate of error. Thus, it

Figure 7.3 Upward streamers from ground near lightning strike points

turns out that the two unconnected streamers associated with the nearer flash are 8 (left) and 10 (right) meters in length. The length of the upward streamer next to the flash on the right measures very nearly 10 meters when allowance is made for the greater distance. The streamers are assumed to be perpendicular to the line of sight in the photograph, and thus these lengths represent lower limits. From the lay of the land viewed in the photograph, this would seem to be a good assumption, with errors of 10 percent or less.

A close examination of the original film revealed a faint downward streamer branching off the nearer channel, at a position corresponding to 18 millimeters up from the ground line on this reproduction. There was also a slight thickening of the main channel at 7 millimeters up. The first position, which translates to 40 meters above ground, may be the *striking distance,* or the height of the last leader step above ground before the onset of the final junction process that initiated the flash. The second position, 15 meters from the ground, probably marks the top of the upward connecting streamer that joined the downward leader. In Figure 7.2 the connection seems to have taken place about 18 meters above the ground — in good agreement with these measurements.

This photograph and a more detailed analysis have been published by Krider and Ladd (1975). *(Photo by Gary Ladd, courtesy of Kitt Peak National Observatory, Tucson.)*

Fine detail in a photograph of two lightning flashes captured on the same film (Figs. 7.4-7.6). The channels show a diffuse halo which may be a corona discharge. Figure 7.4 shows two flashes that were recorded on the same film and within a few seconds of each other. The thunderstorm was about 4 kilometers from the camera. A vertical length of nearly 200 meters is shown in this reproduction, which has been enlarged 3 times. The instrumental and photographic data are the same as for Figure 5.3, in which the spectral response is limited to the region from 6300 Å to 6800 Å, and the radiation from H-alpha is emphasized. However, the film was processed in a phenidone low-contrast developer in order to show fine details in the images over a wide range of illumination. The two main channels appear to be nearly equal in brightness, but the one on the right has a distinct halo, or sheath, associated with it. The channel on the left seems to be slightly fuzzy, but less so than its companion.

If the lack of sharpness or the halo were due to poor focus, or halation in the emulsion and its film base, equally bright channels would appear the same. If one flash were made up of more strokes than the other, and at the same time a crosswind shifted the position of the channel slightly, the one with the greater number of strokes would appear to be thicker, and on the verge of looking like ribbon lightning (see Fig. 3.1). In the presence of a crosswind, only a single-stroke flash would show up as a truly sharp channel in such a photograph. Very sharp and well-defined branches appear over the entire picture. However, just to the right of center is a nearly vertical branch which is an exception: it is fuzzy and becomes fainter toward the top. The appearance of this branch is clear evidence that rain and mist was associated with part of the region covered by the photograph. So perhaps a curtain of rain or mist produced the distinct halo on the right, while the channel on the left was only slightly obscured and diffused.

Figures 7.5 and 7.6 are further enlargements of areas of the previous photograph, centered on the two

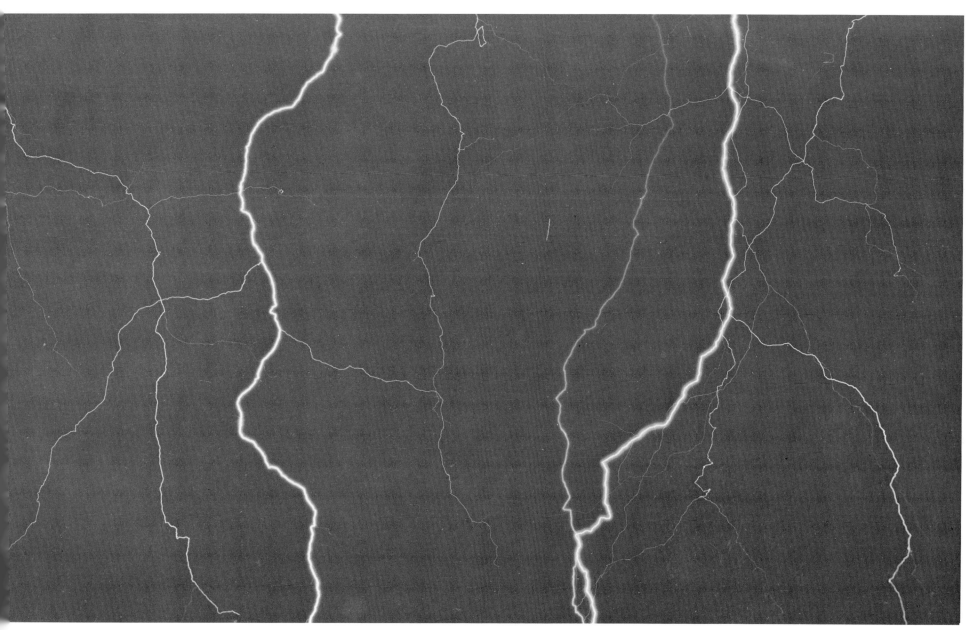

Figure 7.4 Lightning flash showing possible corona discharge from channel

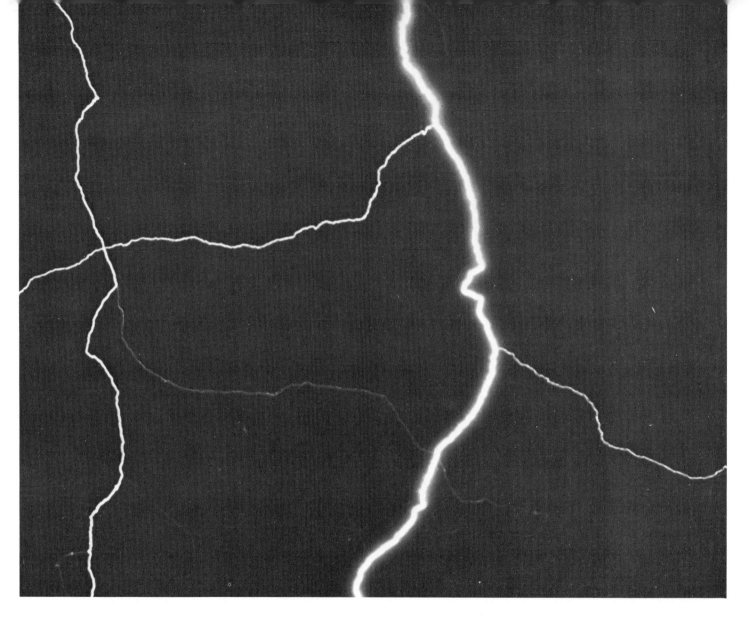

Figure 7.5 Enlargement
showing widened channel

bright flashes on the left and right, respectively. These reproductions were enlarged 10 times from the original negative. Ten millimeters equal about 3 meters at the lightning. The prints were developed in exactly the same way so that shades of light and dark can be compared reliably.

First, note that the image of the bright channel in Figure 7.5 is widened, as already suspected from Figure 7.4. This could be simply an effect of overexposure (brighter flash than its companion), or it could be the effect of several strokes slightly shifted by a crosswind as discussed above. Without a time-streaked photograph,

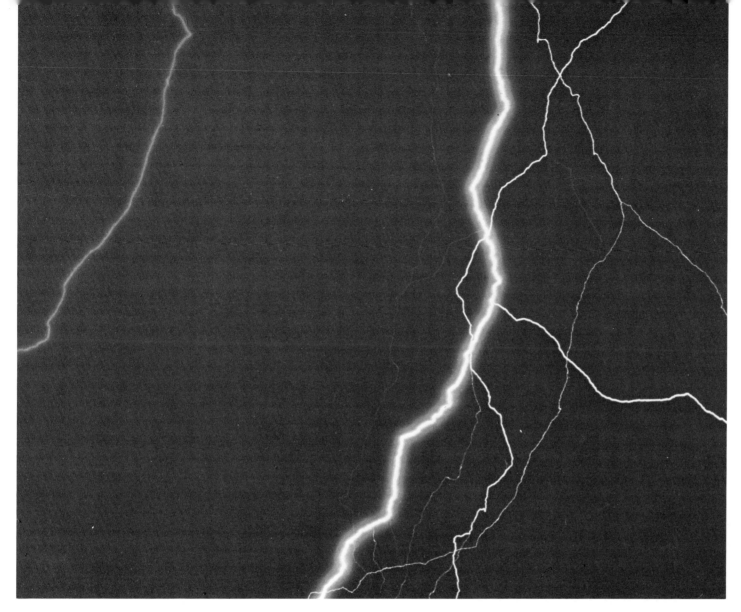

Figure 7.6 Enlargement showing possible corona discharge surrounding channel

or other means of knowing how many strokes made up each flash and how bright they were, the ambiguity cannot be resolved.

Second, notice the short section of a branch in the upper left corner of Figure 7.6. This is part of the fuzzy channel already pointed out in the discussion of Figure 7.4. Comparing its appearance with that of other branches shown in the photograph supports the conclusion that a veil of mist or rain was present (compare with Fig. 5.1). Could a curtain of precipitation have caused a halo to appear along both bright channels, but more conspicuously around one than the other? At first

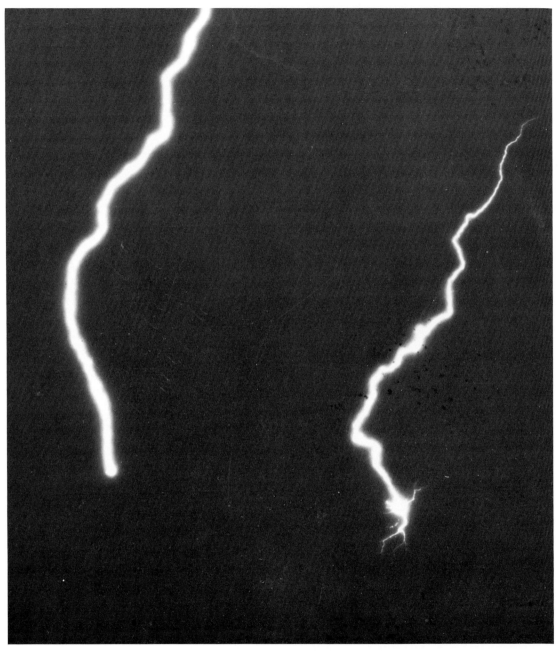

Figure 7.7 Lightning discharges along ground

glance this may seem to be a reasonable explanation but, on closer examination, one sees that the brightest branch connecting with the main channel in Figure 7.6 is sharp and clear. Even with its weaker exposure, it is still surprising that the most prominent branch does not show some evidence of a halo where it connects with the lightning channel. In case something had been lost in making the print, this detail was verified by examining the original negative.

Finally, when one examines the image of the bright channel to the left on the original negative (Fig. 7.5), there is also a faint halo which appears to have the same width as the strong halo associated with the other flash — i.e., about 1.5 meters across. This circumstance makes it even more likely that the halo is something associated with the lightning source rather than being an optical effect in the light path. *(Photos courtesy of Institute of Atmospheric Physics, University of Arizona, Tucson.)*

The possibility of luminous effects arising from lateral corona discharges extending several meters from the main channel has received some attention in studies of the physics of lightning. The possibility of a continuing current luminosity (see Fig. 6.17) relating to such a halo has been suggested, but there is no general agreement as to any connection between them. The reader interested in details can find a short discussion, with technical references, in Hill (1977).

Lightning strikes and splatters over the ground (Figs. 7.7, 7.8). The photograph in Figure 7.7 was taken with a large telephoto camera, focal length 1220 mm, from a distance of approximately 6 kilometers. The camera was pointed toward hilly terrain, but the precise direction is not known, so the ground strike point cannot be identified. However, it seems clear that the flash on the right struck the side of a hill or a large rock. The one on the left probably went behind a ridge, because there is no evidence on the photograph of a ground contact.

The main lightning channel is overexposed and largely obscures an upward streamer which may be located just to the left of the ground contact point. However, several short, tortuous discharge channels can be seen radiating outward from the strike point. The distance scale on the reproduction is approximately 10 millimeters to 6 meters, and, therefore, what appear to be sparks radiating from the end of the lightning flash must have been about 10 meters long. *(Photo courtesy of Institute of Atmospheric Physics, University of Arizona, Tucson.)*

So far as is known to the author, Figure 7.7 shows the only existing photograph of discharges along the ground, taken just at the time the lightning struck. However, there are a number of cases where the effects of such an event have been found and photographed. The next illustration is an example.

On the morning after an intense lightning storm in the vicinity of a golf course about 20 miles southeast of Tucson, a pattern of scorched grass was discovered on the green of the fifth hole, as shown in Figure 7.8. Lightning had struck the fiberglass flagpole without damaging it, and then had proceeded to radiate outward along the surface of the grass in a branched pattern, for distances extending 8 to 12 meters from the cup.

According to local reports, rain had been falling heavily for at least half an hour before lightning struck the course, and it is very probable that the grass was thoroughly wet when the pole was struck. The photograph was taken 3 days after the event, and the area had

Figure 7.8 Lichtenberg figure scorched into putting green by lightning flash

dried out to some extent. The grass was not burned to the degree that it was blackened, but rather it was scorched and turned brown. The result of such a strike depends on the condition of the surface, as well as the peak lightning current and its duration. A comparatively low current continuing for several tenths of a second (a slow burn) may cause more damage than a much higher current lasting for less than a thousandth of a second.

The branching, or dendritic, pattern is characteristic of a phenomenon in laboratory electrical discharges known as the *Lichtenberg figure,* or the *klydonogram.* This fine example of a Lichtenberg figure in nature has been discussed in more detail by Krider (1977). *(Photo by E. Philip Krider, courtesy of University of Arizona, Tucson.)*

Several photographs of Lichtenberg figures, produced both by sparks and by lightning, are contained in *The Lightning Book,* by P. E. Viemeister, along with numerous other illustrations of damage to structures that have been struck by lightning. A more up-to-date, semitechnical book on lightning damage and protection techniques is *Lightning Protection,* by R. H. Golde.

Does a lightning channel change its shape from one stroke to the next? (Fig. 7.9) No two lightning flashes, following different stepped leaders, look the same. Their jagged paths are as varied as the edges of pieces of paper torn from different sheets — they never match up (see Figs. 2.4 and 2.5 for examples). But if two or more strokes follow a succession of dart leaders, the channel seems not to vary — at least not as seen in photographs taken at the distances normally involved in the study of lightning (see Figs. 4.1 and 4.11). When lightning is photographed at close range, with a high-speed framing camera focused on the tip of a nearby tower, there seems to be little, if any, change in the shape of the path of a single stroke from one frame to the next (see Figs. 3.8 and 5.7). However, one combination of experimental conditions has not been mentioned: What about consecutive strokes (separated in time by 20 or 30 milliseconds) photographed at close range with a telephoto lens (which provides a spatial resolution of a few centimeters)? Do the strokes follow the same path when viewed in such scales of time and space?

Figure 7.9, which shows the spectra of two consecutive strokes, seems to answer in the negative: the strokes do not follow the same path. The camera, arranged to work as a slitless spectrograph, was designed to time-resolve the spectrum of a lightning flash. The optics were identical to those used to obtain the spectra in Figure 6.16; the revolving film drum was a modification of the original design. But in this case the spectrograph was focused on a region centered about 6 meters above the tip of a metal tower, at a distance of only 95 meters from the camera lens. The isolated section of lightning channel is 2 meters long, and structures separated by only a few centimeters can be resolved. This particular exposure seems to have been slightly out of focus, so the soft, fuzzy appearance of the spectrum lines should be regarded as instrumental in origin rather than intrinsic to the lightning channel.

These strokes are the first pair in a series of four recorded on the film. Only a short section at the violet end of the spectrum is reproduced here to show the contour of the channel, especially well represented by the bright N II emission line at 3995 Å. The time interval between the strokes is 25 milliseconds.

In an effort to determine the apparent horizontal displacements at corresponding points along the channel, that is, on No. 2 with respect to No. 1, tracings of both were made on transparent cross-section paper. These tracings were then superimposed at their end points, and lateral displacements were read off at 21 equispaced intermediate points. The average displacement, expressed in distance units across the lightning channel and in the plane of the photographs is about 5 centimeters. Multiplying this by the reciprocal of the mean cosine projection factor, i.e., by $\pi/2$, gives 8 centimeters for the estimated average displacement in three dimensions. However, the axis of the camera was pointed upward at an angle of 39 degrees, thus introducing another correction for projection, i.e., secant 39° = 1.28. This correction could be made in an estimate of the average displacement if the channel were vertical, or if it were a straight line whose orientation in altitude and azimuth were known. Since neither the linearity nor the orientation of even such a short section can be assumed, we can conclude only that the lightning flash did not take exactly the same path from one stroke to the next, and that a typical lateral deviation was roughly 10 centimeters, with an uncertainty of perhaps ±50 percent. *(Photo courtesy of Institute of Atmospheric Physics, University of Arizona, Tucson.)*

This photograph has been reproduced in connection with a discussion of apparent changes in the shapes of lightning channels by Salanave (1969*b*).

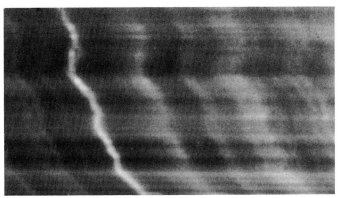

No. 2

Figure 7.9 Slitless spectra of two consecutive strokes showing their different paths

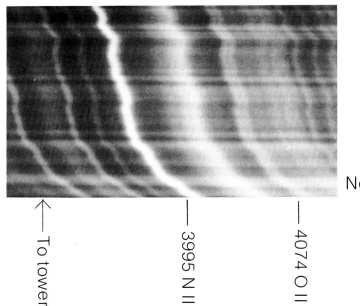

No. 1

To tower   3995 N II   4074 O II

GLOSSARY

This brief collection of words and phrases is limited to those terms that a reader not familiar with the specialized vocabulary used in the description of lightning and its related science and technology might come upon in the course of glancing through this book. Most of these terms are defined or explained in the first place where they are used, but the browser may come upon them elsewhere. On such occasions, this glossary should be helpful.

air discharge — a lightning channel that propagates outward from a cloud but does not reach the ground. It exhibits branching but no return stroke. (See Figs. 2.8, 2.9.)

angstrom — a unit used in the measurement of wavelengths of radiation in the electromagnetic spectrum, usually throughout the range of visible light, and for shorter wavelengths beginning with the ultraviolet and extending toward the region of x-rays. One angstrom unit, symbol Å or A, equals 10^{-10} meter. The visible wavelengths extend from approximately 4000 to 7000 Å.

The angstrom has been almost replaced in physics by the nanometer, symbol nm. 1 nm = 10^{-9} meter or 10 angstroms. Green light, near the middle of the visible spectrum, has a wavelength of about 550 nm.

blazed spectrum — in the application of *diffraction gratings* (see below) to spectrum photography it is possible to design and produce gratings with grooves shaped so that an increased fraction of the incident light goes into the spectrum of a particular order — usually the first on one side of the zero order, that is, the undispersed image of the source.

channel (of a lightning stroke) — the path of the electrical discharge between two clouds, a cloud and the ground, or within a cloud. Under the conditions of electrical stress developed near or within a thundercloud, air becomes *ionized* (see below) and, therefore, a conductor of electricity. This allows the discharge to take place along a tortuous, meandering path which may be as small as a few millimeters, or as much as a few centimeters in diameter.

*cloud discharge (*or *intracloud lightning)* — a lightning discharge between positive and negative charge centers, both within the same cloud.

*cloud-to-cloud discharge (*or *intercloud lightning)* — a lightning discharge between a positive charge center in one cloud and a negative charge center in another cloud; much less common than intracloud discharges, with which they should not be confused. (See Fig. 2.10.)

continuing current — an electrical current that sometimes flows along a lightning discharge channel during the time between discrete strokes in a flash, or after the last stroke. This current may amount to one or two hundred amperes and last for a few hundred milliseconds. It maintains a weak state of ionization (low conductivity) in the channel, together with a faint radiance given off by the heated air. The continuing current is typically followed by a dart leader-return stroke sequence, but in some cases the flash is terminated by it. (See Figs. 4.1 and 4.5.)

continuing current luminosity — a low level of radiance in a lightning channel, occurring during the passage of a current between two return strokes or at the end of a lightning flash. (See Figs. 4.1, 4.3, 4.5.)

corona discharge — a flow of electricity outward from a conductor in a sufficient voltage gradient with respect to its surroundings to ionize the gas and produce a glow. The voltage gradient is not, however, sufficient to produce a complete spark or arc. St. Elmo's fire, often seen at the tip of masts and in the rigging of ships under a highly charged thundercloud, is an example of corona discharge.

coulomb — a unit of quantity of electrical charge. A current of one ampere flowing for one second equals one coulomb of charge, as would 1/10 ampere for 10 seconds, 1000 amperes for 1/1000 second, etc.

cumulonimbus — the ultimate development of the cumulus or heaped cloud. It always develops from a cumulus congestus but differs from it principally in the fact that it more often yields heavy precipitation and usually produces lightning. It is often associated with hail and the development of a plume or anvil. (See Fig. 1.2.)

cumulus congestus — a dense cloud with strong vertical development, characterized by its cauliflower or tower appearance and large size. It may produce abundant precipitation but not lightning activity. (See Fig. 1.1.) It frequently develops into a *cumulonimbus* (see above).

dart leader — a smoothly advancing column of electrical charge that initiates the second and subsequent strokes in a multiple-stroked lightning flash. Dart leaders propagate approximately ten times as rapidly as stepped leaders, because residual ionization in the path of the preceding return stroke maintains a higher conductivity of the air. Compare *stepped leader* (see below). (See Fig. 4.10.)

diffraction grating — an optical device consisting of very fine parallel grooves ruled on metal (for reflection) or on glass (for transmission). Light falling on or passing through such a device is dispersed into its component wavelengths (colors) and can be viewed as a spectrum. There are several spectra arranged in pairs, or orders, on both sides of a ray representing the zero order or line of no dispersal into a spectrum. The first-order spectra each contain the same amount of light unless the grating has been *blazed* (see above).

dodging — a technique used in photographic printing to compensate for extremes of density in the negative. This is usually done with some type of mask, but a finger or a strip of paper may suffice in lieu of anything cut out especially for the purpose. The mask is held above some over-illuminated area on the printing easel and moved back and forth rapidly during the exposure. Dodging is an art rather than an exact procedure.

fish-eye lens — a photographic lens capable of imaging approximately a hemisphere onto a single frame of film in a camera. The full field angle may be anything from 140 to 220 degrees, depending on the design.

halation — a photographic effect which spreads out the image of an otherwise sharp point or line; it is especially noticeable in cases of overexposure. *Diffuse halation* is due to scattering of light in the layer of emulsion. *Reflex halation* is caused by light penetrating the emulsion layer and returning to it by internal reflection from the back surface of the film or glass plate. Reflex halation is practically eliminated in modern films and plates by a special coating, called "backing." (See Figs. 2.1, 2.14.)

ionization — a condition (commonly occurring in a gas under electrical stress or at high temperature) in which specific atoms or groups of atoms lose one or more electrons. Such atoms are referred to as once, twice, or more times ionized; otherwise they are said to be neutral. In physics the roman numeral I designates a neutral atom; II is once-ionized, etc. For example, O I is neutral oxygen, N II is once-ionized nitrogen.

magnetic pinch effect — if a heavy current flows across a spark gap, the luminous discharge may take the form of a cylinder, or sheath, whose diameter varies periodically along its length. That is, the column of plasma will appear to be pinched at several places along the spark; hence the analogy to a string of sausages in one theory on the origin of bead lightning. (See Fig. 3.7.)

return stroke (or *ground flash*) — the intensely luminous lightning streamer which propagates upward from the ground to the base of a cloud. It may follow either a *stepped leader* or a *dart leader* (which see). One or more consecutive strokes constitute a lightning flash. The current in a return stroke may be 10 to 20 kiloamperes or higher, but lasts only on the order of 50 microseconds. (See Figs. 2.1, 2.2.)

slitless spectrum — a display of a range of wavelengths emitted by a distant source of light, where the narrow, linelike geometry of the source eliminates the need for the collimator lens and the slit that are required in the usual laboratory spectroscope. (See Figs. 6.1, 6.3.)

Stark effect — the splitting or broadening of certain spectrum lines, particularly those of hydrogen and helium when the source atoms are in a strong electric field. The field may be applied externally, as in a cathode ray tube, in which case the affected emission lines are split into several components which can be resolved with a slit spectrograph and high dispersion.

In the case of an ionized gas at high temperature, as in a lightning stroke, the electric field is applied internally by ions and free electrons near the affected atoms. Since these particles are in rapid, random motion the fields are very inhomogeneous and the spectrum lines are smeared or broadened, rather than split into distinct components.

stepped leader — an intermittently advancing column of electrical charge (ionized air) which establishes the path, or channel, for a subsequent lightning stroke. In the most commonly observed case of a flash between cloud and ground, the leader moves downward in sections (steps), each about 50 meters long, and carries a negative charge. (See Fig. 4.6.)

telluric bands — complex structures in the spectrum of light that has passed through the earth's atmosphere, whether along a path directed downward through its entire depth, as with sunlight, or for only a few kilometers along the ground, as in observations of lightning. These spectral bands are particularly conspicuous in the deep red and infrared, at which wavelengths oxygen and water vapor are the strongest absorbers. (See Fig. 6.10.)

upward connecting streamer — a positively charged spark channel that propagates several tens of meters up from the ground to contact the negative stepped leader that will initiate the first return stroke in a lightning flash. It may be thought of as the discharge that short-circuits the electrical charge from the leader to the ground — rather like connecting both terminals of a storage battery with a heavy wire. (See Figs. 7.2, 7.3.) The word *streamer* has several connotations in discussions of lightning discharge theory; these should not be confused with the specific description given here.

BIBLIOGRAPHY

REFERENCES

Anderson, R. V.; Bjornsson, S.; Blanchard, D. C.; Gathman, S.; Hughes, J.; Jónasson, S.; Moore, C. B.; Survilas, H. J.; and Vonnegut, B. 1965. Electricity in volcanic clouds. *Science* 148: 1179-89.

Berger, K. 1966. Photographische Blitzuntersuchen der Jahre 1955-1965 auf dem Monte San Salvatore. *Bull. Schweiz. Electrotech. Ver.* 57: 1-22.

_____. 1967. Novel observations on lightning discharges. *J. Franklin Institute* 283: 478-525.

_____. 1972. Methoden und Resultate der Blitzforschung auf dem Monte San Salvatore bei Lugano in den Jahren 1963-1971. *Bull. Schweiz. Electrotech. Ver.* 63: 1403-22.

_____. 1973. Ozillographische Messungen des Feldverlaufs in der Nähe des Blitzeinschlags auf dem Monte San Salvatore. *Bull. Schweiz. Electrotech. Ver.* 64: 120-36.

Berger, K., and Vogelsanger, E. 1969. New results of lightning observations. In *Planetary electrodynamics,* ed. S. C. Coroniti and J. Hughes, vol. 1, pp. 489-510. New York: Gordon and Breach.

Brook, M.; Moore, C. B.; and Sigurgeirsson, T. 1974. Lightning in volcanic clouds. *J. Geophys. Res.* 79: 472-75.

Evans, W. H., and Walker, R. L. 1963. High-speed photographs of lightning at close range. *J. Geophys. Res.* 68: 4455-61.

Golde, R. H. 1967. The lightning conductor. *J. Franklin Institute* 283: 451-77.

Hagenguth, J. H., and Anderson, J. G. 1952. Lightning to the Empire State Building. *Trans. AIEE* 71 (Part 3): 641-49.

Hill, R. D. 1973. Lightning induced by nuclear bursts. *J. Geophys. Res.* 78: 6355-58.

————. 1977. Anomalous behavior of H lines in lightning spectra. In *Electrical processes in atmospheres,* ed. H. Dolezalek and R. Reiter, pp. 647-51. Darmstadt, W. Germany: Dietrich Steinkopff Verlag.

Krider, E. P. 1966. Comment on paper by Leon E. Salanave and Marx Brook, "Lightning photography and counting in daylight, using H-alpha emission." *J. Geophys. Res.* 71: 675.

————. 1974. An unusual photograph of an air lightning discharge. *Weather* 29: front cover, 24-27.

————. 1977. On lightning damage to a golf course green. *Weatherwise* 30: front cover, 111.

Krider, E. P., and Ladd, C. G. 1975. Upward streamers in lightning discharges to mountainous terrain. *Weather* 30: 77-81.

McEachron, K. B. 1941. Lightning to the Empire State Building. *Trans. AIEE* 60: 885-89.

Meinel, A. B., and Salanave, L. E. 1964. N_2^+ emission in lightning. *J. Atmos. Sci.* 21: 157-60.

Moore, C. B., and Vonnegut, B. 1977. The Thundercloud. In *Lightning,* ed. R. H. Golde, vol. 1, 92. New York: Academic Press.

Moore, C. E. 1945. A multiplet table of astrophysical interest. *Contrib. Princeton Univ. Obs.* No. 20.

Newman, M. M. 1969. Lightning discharge simulation and triggered lightning. In *Planetary electrodynamics,* ed. S. C. Coroniti and J. Hughes, vol. 2, pp. 213-19. New York: Gordon and Breach.

Newman, M. M.; Stahmann, J. R.; Robb, J. D.; Lewis, E. A.; Martin, S. G.; and Zinn, S. V. 1967. Triggered lightning strokes at very close range. *J. Geophys. Res.* 72: 4761-64.

Ogawa, T., and Brook, M. 1964. The mechanism of the intracloud lightning discharge. *J. Geophys. Res.* 69: 5141-50.

Orville, R. E. 1966. High-speed, time-resolved spectrum of a lightning stroke. *Science* 151: 451.

————. 1968a. A high-speed, time-resolved spectroscopic study of the lightning return stroke. *J. Atmos. Sci.* 25: 827-56.

————. 1968b. Photograph of a close lightning flash. *Science* 162: 666-67.

————. 1977a. Wind profile in the sub-cloud layer of a thunderstorm. *J. Geophys. Res.* 82: 3453-56.

————. 1977b. Lightning spectroscopy. In *Lightning,* ed. R. H. Golde, vol. 1, chap. 8, New York: Academic Press.

Orville, R. E., and Berger, K. 1973. An unusual lightning flash initiated by an upward-propagating leader. *J. Geophys. Res.* 78: 4520-25.

Orville, R. E.; Helsdon, J. H., Jr.; and Evans, W. H. 1974. Quantitative analysis of a lightning return stroke for diameter and luminosity changes as a function of space and time. *J. Geophys. Res.* 79: 4059-67.

Orville, R. E., and Salanave, L. E. 1970. Lightning spectroscopy — photographic techniques. *Applied Optics* 9: 1775-81.

Prueitt, M. L. 1963. The excitation temperature of lightning. *J. Geophys. Res.* 68: 803-11.

Salanave, L. E. 1961. The optical spectrum of lightning. *Science* 134: 1395-99.

————. 1964. The optical spectrum of lightning. In *Advances in geophysics,* ed. H. E. Landsberg and J. Van Mieghem, vol. 10, pp. 83-98. New York: Academic Press.

————. 1969*a.* Recent advances in the observations of lightning spectra. In *Planetary electrodynamics,* ed. S. Coroniti and J. Hughes, vol. 1, pp. 449-66. New York: Gordon and Breach.

————. 1969*b.* Discussion of a paper by M. M. Newman. In *Planetary electrodynamics,* ed. S. Coroniti and J. Hughes, vol. 2, pp. 220-21. New York: Gordon and Breach.

Salanave, L. E., and Brook, M. 1965. Lightning photography and counting in daylight, using H-alpha emission. *J. Geophys. Res.* 70: 1285-89.

Salanave, L. E.; Orville, R. E.; and Richards, C. N. 1962. Slitless spectra of lightning in the region from 3850 to 6900 angstroms. *J. Geophys. Res.* 67: 1877-84.

Thorarinsson, S. 1967. *Surtsey, the New Island in the North Atlantic.* Translated by Sölvi Eyesteinsson. New York: Viking Press.

Uman, M. A. 1969*a.* *Lightning.* New York: McGraw-Hill.

————. 1969*b.* Decaying lightning channels. In *Planetary electrodynamics,* ed. S. C. Coroniti and J. Hughes, vol. 2, pp. 199-209. New York: Gordon and Breach.

————. 1971. Comparison of lightning and a long laboratory spark. *Proc. I.E.E.E.* 59: 457-66.

Uman, M. A.; Seacord, D. F.; Price, G. H.; and Pierce, E. T. 1972. Lightning induced by thermonuclear detonations. *J. Geophys. Res.* 77: 1591-96.

Workman, E. J.; Brook, M.; and Kitagawa, N. 1960. Lightning and charge storage. *J. Geophys. Res.* 65: 1513-17.

————. 1962*a.* Continuing currents in cloud-to-ground lightning discharges. *J. Geophys. Res.* 67: 637-47.

————. 1962*b.* Quantitative study of strokes and continuing currents in lightning discharges to ground. *J. Geophys. Res.* 67: 649-59.

READING LIST

Note: The following list is highly selective. It is intended to include just those articles, books, and periodicals that any persons seeking additional information should look into — at the level of technicality appropriate to their interest, as indicated in the divisions of the following list. For the research-oriented reader, the technical sources (listed in order of least to most specialized) will give a rather complete treatment of methods and results and, via the reference lists they contain together with the periodical literature sources listed, will lead to a thorough review of the development and the present state of knowledge about lightning and related aspects of atmospheric electricity.

Nontechnical Sources

Battan, Louis J. *The Thunderstorm.* New York: New American Library, 1964.

Encyclopaedia Britannica, 15th ed., 1974. "Lightning." A comprehensive, well-illustrated review by R. E. Orville. The 14th edition (1969) contains an article by E. T. Pierce, to which the reader is referred for a treatment of the subject from a somewhat different perspective.

Frazier, Kendrick. *The Violent Face of Nature.* New York: Morrow, 1979. Chapters 1, 2, and 3.

McGraw-Hill Encyclopaedia of Science and Technology. 4th ed., 1977. "Lightning." General description by L. J. Battan; optical spectrum by L. E. Salanave.

Schonland, Basil F. J. *Flight of the Thunderbolts.* 2nd ed. Oxford: Clarendon Press, 1964.

Singer, Stanley. *The Nature of Ball Lightning.* New York: Plenum Press, 1971.

Uman, Martin A. *Understanding Lightning.* Carnegie, Pa.: Bek Technical Publications.

Viemeister, Peter E. *The Lightning Book.* Garden City, N. Y.: Doubleday, 1961. Reprinted in paperback by M.I.T. Press, Cambridge, Mass., 1972.

Technical Sources

Uman, Martin A. *Lightning.* New York: McGraw-Hill, 1969.

Golde, Rudolf H., ed. *Lightning.* 2 vols. New York: Academic Press, 1977.

Coroniti, Samuel C., ed. *Problems of Atmospheric and Space Electricity.* Amsterdam: Elsevier Press, 1965. This is the publication of the proceedings of the 3rd International Conference on Atmospheric Electricity, held 5-10 May 1963 in Montreux, Switzerland.

Coroniti, Samuel C., and Hughes, James, eds. *Planetary Electrodynamics.* 2 vols. New York: Gordon and Breach, 1969. This is the publication of the proceedings of the 4th International Conference on Atmospheric Electricity, held 12-18 May 1968 in Tokyo, Japan.

Dolezalek, Hans, and Reiter, Reinhold, eds. *Electrical Processes in Atmospheres.* Darmstadt, W. Germany: Dietrich Steinkopff Verlag, 1977. This is the publication of the proceedings of the 5th International Conference on Atmospheric Electricity, held 2-7 September 1974 in Garmisch-Partkirchen, West Germany.

Golde, Rudolf H. *Lightning Protection.* London: Edward Arnold Publishers, 1973.

Journals

Journal of Geophysical Research (JGR), published monthly by the American Geophysical Union, Washington, D.C., is divided by category into several separately bound sections. "Oceans and Atmosphere" ("The Green Section") is currently the most active in publishing results about lightning and atmospheric electricity.

Journal of the Atmospheric Sciences (JAS), published monthly by the American Meteorological Society, Boston, occasionally contains papers on atmospheric electricity, particularly as they may relate to general meteorology.

Journal of Atmospheric and Terrestrial Physics (JATP), published monthly by Pergamon Press, London, is approximately equivalent to *JGR,* for Great Britain.

Quarterly Journal of the Royal Meteorological Society (Quart. J. Royal Met. Soc.), published quarterly by the Royal Meteorological Society, Bracknell, England, is another source of fundamental research papers on cloud physics, especially those concerned with precipitation and electrification.

PHOTO CREDITS

(Figure numbers follow contributors' names)

Hugo Binz, Baden, Switzerland	2.6	New Mexico Institute of Mining and	2.12, 2.13, 4.1,
Brookhaven National Laboratory, Long Island	3.3	Technology, Socorro (Marx Brook)	4.3-4.5, 4.11, 4.12,
Rudolf H. Golde, London	3.4		4.20
Harvard College Observatory	6.1	Richard E. Orville, State University	2.15, 2.16,
Institute for High-Tension Research,		of New York at Albany	5.8, 6.18, 6.19,
University of Uppsala, Sweden			6.21b, 7.2
(Stig Lundquist, Victor Skuka)	6.20	Swiss High Voltage Research	2.5, 2.8,
Sigurgeir Jónasson, Vestmannaeyjar, Iceland	1.5, 3.9, 3.10	Committee, Zurich (Karl Berger,	2.14,
E. Philip Krider, Institute of		Ernst Vogelsanger, Hugo Binz)	4.6-4.9, 4.13-4.17
Atmospheric Physics, University of Arizona	7.8	William L. Taylor, National Severe	
Gary Ladd, Kitt Peak National Observatory	7.3	Storms Laboratory, Norman, Oklahoma	4.2
Lightning and Transients Research Institute,		U.S. Naval Ordnance Laboratory,	
Miami (James R. Stahmann)	2.19	Silver Spring, Maryland	3.6, 3.7
Los Alamos Scientific Laboratory		University of Arizona Institute of	1.2, 2.1, 2.2, 2.4,
(Frank Berry, William Regan)	3.11	Atmospheric Physics (Louis Battan,	2.9-2.11, 3.2, 3.5,
Lowell Observatory	6.2	Walter Evans, E. Philip Krider, Clyde	3.8, 4.10,
George Marcek, Tucson	1.1, 1.3, 1.4, 2.3,	Richards, Leon Salanave)	4.21-4.23, 5.1-5.7,
	2.7, 2.18a, 2.20-2.22,		6.3-6.17, 6.21a,
	3.1, 4.18, 4.19		7.4-7.7, 7.9
National Aeronautics and Space Administration,		Jouko Wesanterä, Helsinki	7.1
Johnson Space Flight			
Center (Donald Arabian)	2.17, 2.18b		

INDEX